Strength in Numbers

Strength in Numbers

Discovering the Joy and Power of Mathematics in Everyday Life

Sherman K. Stein

John Wiley & Sons, Inc.

New York • Chichester • Brisbane • Toronto • Singapore

This text is printed on acid-free paper.

This publication is designed to provide accurate and authoritative information in regard to the subject matter covered. It is sold with the understanding that the publisher is not engaged in rendering legal, accounting, or other professional services. If legal advice or other expert assistance is required, the services of a competent professional person should be sought.

Library of Congress Cataloging-in-Publication Data:
Stein, Sherman K.
    Strength in numbers: discovering the joy and power of mathematics in everyday life / Sherman K. Stein.
       p.   cm.
    Includes bibliographical references (p.   –   ) and index.
    ISBN 0-471-15252-8 (cloth : alk. paper)
    1. Mathematics—Popular works.   I. Title.
QA93.S684 1996
510—dc20   95-48056

Printed in the United States of America
10 9 8 7 6 5 4 3 2 1

You and I—all of us—can explore the inner and outer worlds far more than we imagine possible. Many of us close the doors too soon. I dedicate this book to all who are willing to open closed doors and open even wider the doors already open.

CONTENTS

Contents

ACKNOWLEDGMENTS

As the manuscript of this book went through several revisions, it bene-fited from the comments of a wide variety of critics: professors of mathe-matics, laypersons, and junior high and high school students. It is a pleasure to acknowledge their assistance in making this exposition acces-sible to a broad audience.

Chris Garrett's beginning algebra class at Emerson Junior High School in Davis, California, and Joanne Moldenhauer's analysis class at Davis Senior High School commented on the mathematics chapters as "extra credit" assignments. I was impressed by the students' insights and the clar-ity with which they expressed them.

Pat King and Heather Wade, mathematics teachers at Holmes Junior High School in Davis, offered several suggestions concerning the mathe-matics chapters and the chapters on jobs and reform.

Don Chakerian, professor of mathematics at the University of California at Davis, Sanford Siegel, professor of mathematics at the University of Rochester, and Lester Lange, professor of mathematics at California State University at San Jose, read the entire manuscript and caught several slips (even the introduction of a new false myth). Don Albers, head of publications at the Mathematical Association of America, Henry Alder, professor of mathematics at the University of California at Davis, Anthony Barcellos, professor of mathematics at American River College, and Anthony Wexler, professor of mechanical engineering at the University of Delaware, severe critics all, forced me to strengthen my case in the more polemical chapters. Sharon Dugdale, professor of mathematics education at the University of California at Davis, and Elaine Kasimatis, professor of mathematics at California State University at Sacramento, substantially improved the two chapters on reforms.

My wife, Hannah Stein, was the first person to read each chapter, often in its "pen-and-ink" stage. Her many suggestions helped clarify and humanize the exposition, for she insisted that the book not read like a mathematics text. No words can adequately acknowledge her contribution to this book any more than they can to my life.

Since I wanted the book to be accessible to readers who might have been away from mathematics a long time, or might not have used much mathematics in their careers, I also asked my friends, neighbors, relatives, and children to read various chapters. Their comments, which forced me to work much harder and longer than I had planned, made me meet my readers not halfway, but all the way. These critics—Ted Gould, Paul Jacobs, Dan Keller, Jane Keller, Don Kunitz, Allegra Silberstein, Lori Snyder, Joshua Stein, and Susanna Stein—came from such varied careers as library work, elementary through university teaching, journalism, law, marine biology, and public policy.

The reference librarians at the University of California at Davis made the millions of books and thousands of periodicals in its library as accessible as the books on my living room shelves. My thanks go to Rafaela Castro, Patsy Inouye, Linda Kennedy, Jane Kimball, Sandra Lamprecht, David Lundquist, Marcia Meister, Opritsa Popa, Juri Stratford, and Michael Winter.

I also wish to express my appreciation to my friend Roland Hoermann, professor of German, who translated Moritz Cantor's discussion of the possible use of the 3–4–5 right triangle by the Egyptians.

Anthony Barcellos created the illustrations, for which I am deeply appreciative.

I hope that anyone who has suggestions for improving the book will send them to me at the Mathematics Department, University of California at Davis, Davis, CA 95616-8633. (e-mail: stein@math.ucdavis.edu)

GUIDE TO
CHAPTER ORDER

Part I can be read in any order, though Chapters 2 and 3 form a natural unit, as do Chapters 9 and 10, and Chapters 11, 12, and 13.

In Part 2, Chapter 14, which advises how to read the language of mathematics, is central. Chapters 18 and 19 form a unit, and part of Chapter 25 uses Chapters 17 and 24.

In Part 3, Chapters 27 through 31 should be read in order after Chapters 18 and 24. Chapter 32 depends mainly on Chapter 18.

PART ONE

About Mathematics

··· 1 ···

The Many Faces of Mathematics

Practically everyone can understand and enjoy mathematics and appreciate its role in modern society. More generally, I feel that we develop only a small part of our potential, not only in mathematics but also in art, carpentry, cooking, drawing, singing, and so on. We close up too soon. Each of us can reach a higher level than we imagine if we are willing to explore the world and ourselves. I hope that this book will help people explore and feel at home in the world of mathematics.

Young people should be aware that mathematics is a tool in many trades and professions. About two out of every three better-paying jobs require mathematics beyond arithmetic, either as part of the training or for day-to-day use. Only about one out of ten of the lower-paying jobs has such a requirement. (I base these figures on the data in Chapter 10.) This contrast shows how important mathematics is in a highly technologized economy. The more mathematics a person knows, the more doors to careers are open. Because mathematics is such a bread-and-butter issue, I devote an entire chapter—Chapter 10—to describe what mathematics is needed (or is not) in many occupations. (Chapter 9 also concerns this topic.)

It is vital that parents encourage their children to study mathematics. And everyone should be less intimidated when "experts" use numbers and computers to influence decisions. (Chapters 3 and 4 offer ways to protect yourself from this common form of number abuse.)

The purpose of this book is to spread the gospel of mathematics, to carry the word to unbelievers and believers alike. I hope to bring back into the fold those who were turned off by an unpleasant experience in school (usually by age 12) or who simply drifted away. Also, by presenting fresh examples, chosen either for their beauty or for their practical importance, I hope to reinforce the affection of those who have had only happy encounters with mathematics.

Mathematics is like the proverbial elephant as described by three blind persons. Upon touching a leg, one says, "It's like a tree." Another, touching the trunk, reports, "It's like a snake." The third, touching an ear, says, "It's like a bat."

So it is with mathematics. If you know it only as a tool for doing arithmetic—finding lengths and areas and calculating costs and profits—it resembles a hammer or a screwdriver. If you see it used to describe gravity or the geometry of chromosomes, you might think of it as the language of creation for the physical and biological universe. Or, taking a course in geometry or calculus, you might call mathematics a means to develop analytical skills, a training ground for careers such as business, law, or medicine. This last is the view of Conrad Hilton, founder of the hotel chain, as expressed in his book *Be My Guest:*

> I'm not out to convince anyone that calculus, or even algebra and geometry, are necessities in the hotel business. But I will argue long and loud that they are not useless ornaments pinned onto an average man's education. For me, at any rate, the ability to formulate quickly, to resolve any problem into its simplest, clearest form has been exceedingly useful. It is true that you don't use algebraic formulae but . . . I found higher mathematics the best possible exercise for developing the mental muscles necessary to this process . . . .
>
> A thorough training in the mental disciplines of mathematics precludes any tendency to be fuzzy, to be misled by red herrings. . . .

Hilton is not the only businessman to emphasize the importance of mathematics. Charles Munger, a partner of the legendary financier Warren Buffett, in a talk at the University of Southern California in 1994 advised the business students,

> First there's mathematics. You've got to be able to handle numbers—basic arithmetic. And the great useful model, after compound interest, is the elementary math of permutations and combinations. It's very simple algebra. It's not hard to learn. What is hard is to get so you use it routinely almost every day.
>
> At Harvard business school the thing that bonds the first-year class together is decision-tree theory. They take algebra and apply it to real life problems. Students are amazed to find that high school algebra works in life.

There are still other ways to touch the elephant of mathematics. If you experience the beauty of some of its discoveries and arguments, you may view mathematics as an art form, like music or painting. If you think about its tantalizing but unanswered questions, you might even compare it to an unexplored region of the earth. All of these descriptions are valid, but none tells the whole story.

If you touched mathematics mainly in long lists of dreary calculations, assigned at the rate of a page a day, or mindless rules for which no explanations were given, you had a most unfortunate perspective. In this case, you might have seen mathematics as a punishment and blamed yourself, when you had a right to blame your teacher or your text.

The way a subject is taught certainly influences how well a pupil learns it. Consider pupils who, flunking French, conclude, "I'm hopeless. I just don't have a talent for languages." They completely overlook that they have already mastered English. If those students were to spend a year or two in France, they very likely would learn to speak French fluently, even well enough to earn an A+ back home. The grade a teacher assigns does not measure the pupil but rather how well the pupil performed under a particular mode of instruction. In a sense, it evaluates the teacher, too.

Of all subjects, mathematics can be the best taught, or the worst. In mathematics, all the cards can be put on the table: Nothing has to be taken on faith or on the say-so of some authority figure. Everything should make sense. Guided by a well-prepared teacher, students can conduct their own experiments, make their own discoveries, and uncover many of the basic principles without being told. These experiments require no fancy equipment. Pencil and paper, a calculator, a ruler, a piece of string, dice, and pennies will do.

Learning mathematics is quite a contrast to, say, learning physics, biology, or history. When studying the structure of the atom or the anatomy of the cell, the student must depend on the assertions of countless physicists and biologists. As for the study of history, we have

The past,
hidden under layer
upon layer,
upon layer,
of historians.

Nothing need come between a student and the concepts of mathematics.

How can mathematics be the worst taught? Because it can be presented as a collection of procedures to calculate numbers, which may be of no interest or use. With each new page of drill, the pupil grows more alienated. Eventually, when the pupil hears an oracular pronouncement as, for example, "To divide by a fraction, you turn it upside down and multiply," or, "A negative times a negative is positive," the alienation becomes complete. (Simple explanations of these mysteries are to be found in Chapters 20 and 25.)

In the United States, a person who doesn't do well in mathematics often decides, "I just don't have mathematical talent." The prevailing wisdom has it that the ability to do mathematics lies in the genes and that no amount of work will make a difference, as though it were like being left-handed or having perfect musical pitch. In other countries, the belief is that if you work at mathematics, you will master it. There, mathematics is treated in the same practical manner as carpentry or swimming: If you practice, you improve. High expectations breed high performance. The guiding principle elsewhere is "Teach it right the first time." This means you don't have to spend time later "reviewing" what wasn't learned in the first place.

What John Galbraith wrote about money applies to mathematics also. Just replace the word *money* with the word *mathematics* throughout the following quote.

> Those who talk of money and teach about it and make their living by it gain prestige from cultivating the belief that they . . . have insights that are nowise available to the ordinary person. Though professionally rewarding and personally profitable, this . . . is a . . . form of fraud. There is nothing about money that cannot be understood by the person of reasonable curiosity, diligence, and intelligence. There is nothing in the following pages that cannot be so understood.

In his book *The Blind Watchmaker*, biologist Richard Dawkins reassured readers:

> The important thing to remember about mathematics is not to be frightened. It isn't as difficult as the mathematical priesthood sometimes pretends. Whenever I feel intimidated, I always remember Silvanus Thompson's dictum in *Calculus Made Easy:* "What one fool can do, another can."

Chapter 16 gives readers the opportunity to see that Dawkins is right by solving a problem that embarrassed several mathematicians.

Mathematics splits into two disciplines, almost opposite in spirit. On the one hand, it provides almost mindless ways to carry out routine calculations, now often executed by calculator, cash register, or computer. On the other hand, by providing a language of great precision, mathematics enables us to think about complex decisions in an orderly way, not just by anecdote, guesswork, and persuasive rhetoric.

Our relation to mathematics is like our relation to many parts of our high-tech civilization. We can push keys, move cursors, and turn dials on many gadgets without the faintest idea of how they work. When they break down, we cannot fix them ourselves. The inhabitants of ancient Rome, if dropped into our world, would at first be intimidated and confused, but given a few weeks to adjust, they, too, could push keys, move cursors, and turn dials. Conversely, stripped of the ornaments of our civilization, deprived of the devices we do not understand, we would find that we are not that different from the inhabitants of ancient Rome. Carried back there, we, too, could adjust to their lifestyle.

The more we grasp how the world around us works—in particular, the role of mathematics in our society—the less we will be at the mercy of forces we don't understand. The more we will feel at home, the less we will feel like aliens surrounded by mysteries in our kitchens, in our living rooms, and under the hoods of our cars.

I hope that this book will help us catch up with the thinking of John Adams, the second president of the United States. He wrote his wife, Abigail, in 1780, while in Europe to obtain a treaty of peace:

> I must study politics and war, that my sons may have liberty to study mathematics and philosophy, geography, natural history and naval architecture, navigation, commerce and agriculture, in order to give their children a right to study painting, poetry, music, architecture, statuary, tapestry and porcelain.

Here Adams is speaking of the practical side of mathematics. But, had he been writing to Abigail today, he would place mathematics with the arts as well, as later chapters will show.

This book is divided into three parts, each of a different character.

Part 1, a general commentary, consists of more-or-less independent chapters. Among other things, this section

- shows how we can protect ourselves from the abuse of numbers;
- describes the triumphs and limits of computers;
- debunks some myths about mathematics and mathematicians;

- outlines the surprising applications of mathematics;
- describes the mathematics used in various trades and professions; and
- discusses the reform of mathematics instruction.

Part 2 offers fresh views of concepts often met in school, such as

- the geometric series and its application;
- the theorem of Pythagoras;
- why you turn a fraction upside down to divide by it;
- what π is;
- why a negative times a negative is a positive; and
- how to draw a picture of an equation.

Part 3

- develops a technique for finding some unknown quantity;
- finds the varying steepness of a curve;
- shows how to compute an area bound by a curve; and
- obtains a remarkable connection between a circle and all the odd whole numbers that gives us a way to compute π without drawing a single circle.

The chapters in Part 3 are closely linked. The dependencies among some of the earlier chapters are shown in the guide on page xiii.

Chapter 16 guides readers to discover mathematical ideas on their own and to feel the excitement of exploration. But the arguments behind some mathematical discoveries are so technical that only specialists can follow them. Mathematics, like many fields of knowledge, has splintered into so many subdisciplines that mathematicians themselves cannot read research papers outside their own specialty.

Even so, we can become familiar with a mathematical discovery, though the reasoning behind it will have to stay a tantalizing mystery. For instance, Chapter 15 describes some amazing and profound facts about elementary arithmetic but doesn't show why they are true. I admit that I have not read the arguments myself, for they lie far from my own area of expertise. Still, I can appreciate the discoveries, just as I can appreciate other great accomplishments, such as climbing Mount Everest, or landing on the moon, or painting the Sistine Chapel. In most chapters, the reader will play a role midway between explorer and spectator: that of an active partner. The word *mathematics,* which we usually think of just as a noun, is, in truth, also an active verb.

Anyone "of reasonable curiosity, diligence, and intelligence," who can do arithmetic and who follows the advice in Chapter 14, can master and enjoy any of the chapters in this book. In a few places, beginning algebra will come in handy, but it is not necessary.

After finishing this book, you should have a clearer idea of the importance of mathematics in the "real" world and the ability to read the language of mathematics. I hope, in addition, you will have gained an appreciation of the beauty of mathematics and the elegance of its reasoning.

If I move the reader to view mathematics as Thomas Jefferson did, I will have accomplished my mission. In a letter to a friend in 1811, he wrote, "Having to conduct my grandson through the course of mathematics, I have resumed that study with great avidity. It was ever my favorite one. No uncertainties remain in the mind; all is demonstration and satisfaction." This from a person who made all knowledge, pure and applied, his own.

··· 2 ···

The Spell of Cool Numbers

When I hear that light travels 186,283 miles a second, I believe it. I am impressed that anyone could measure the speed so accurately, but I am assured that scientists with no ax to grind have independently checked it. It is a cool number. By "cool," I mean that it is not part of a serious controversy. Though noncontroversial, it can, as I will show, arouse warm feelings. In the next chapter, I turn to the hot numbers, the kind that play a role in major decisions and tend to be elastic. Those are the dangerous numbers, and I will show how we can protect ourselves from their insidious influence.

Once when I was using the elevator in a modern Manhattan apartment building, I noticed something strange on the control panel. The column of odd numbers abruptly switched to even numbers, with 14 right above 11. "What's going on?" I wondered. Of course, it took only a moment to find the answer: There was no floor numbered 13, as the picture of the panel shows. An ancient superstition had managed to survive in the midst of the most up-to-date, sophisticated technology.

In the building, there was no missing floor. There were 18 floors, and the 13th looked just like the rest. The panel listed the top floor

How do I get to the 13th floor?
(Photo by Donna Binder)

as 19. The superstition concerned the number 13 itself, not the actual floor. A sheer number, something so abstract, so seemingly noncontroversial, so "cool," had managed to arouse an emotion: fear.

Back in 1884, a group of New Yorkers started the Thirteen Club with the goal of putting an end to the 13-superstition. They dined together on the 13th of every month, with 13 at the table, and set their dues at 13 cents per month. Even so, they could report that they were as healthy, prosperous, and as long-lived as members of any other club. In spite of their efforts, 13-phobia persists.

I happen to be especially fond of the number 13. Partly because I like to mock 13-phobia, partly because 13 is the sum of two square numbers, $13 = 4 + 9$. (A square number is one obtained by multiplying a whole number by itself. Since 9 is 3×3, it is a square number.)

But 13 is not my favorite number. Since I was a little boy, 6 has played that role. I think my affection for 6 goes back to my interest in my big brother's antique revolver, a six-shooter. Years later, I discovered that 6 was held in high esteem by the ancient Greek mathematicians, for it equals the sum of its divisors other than itself, $6 = 1 + 2 + 3$. The Greeks called any number that coincides with the sum of its divisors other than itself "perfect." The next perfect number is 28, which equals $1 + 2 + 4 + 7 + 14$. No one knows whether there is a largest perfect number, and no one has ever found an odd perfect number. (It is known that there is no odd perfect number less than the number 1 followed by 300 zeros.) These are just two of the countless "simple" things that mathematicians do not know.

Recently, I have taken a fancy to the neglected number ⅗. When someone asks me for my favorite number, I tell them, "It's ⅗," and wait for their response. Typical is, "Why in heaven's name would you think of that number?" "Well, ½ and ⅔ get a lot of attention. In between them lies ⅗, which no one seems to notice. That's sad, isn't it? Besides, 3 and 5 are rather pleasant numbers."

I mention ⅗ not just to make it feel appreciated. My real purpose is to find out how people feel about fractions. Usually they do not have happy memories. To develop a more positive attitude, I include Chapter 20 in this book. In a few pages, it explains all that anyone ever needs to know about these perfectly decent numbers.

If I have the right to admire the number 6, then I should not be surprised that some people fear the number 13. My admiration and their fear both show that a number can, like a word, carry overtones and connotations. I'll give a few examples where a number has leapt out of the number system to take on a life of its own.

Half a century ago, the mile run practically turned the number 4 into an impenetrable brick wall, as real an obstacle as the sound barrier seemed to be to the speed of airplanes.

In 1945, Gunder Hägg ran the mile in 4 minutes 1.4 seconds and held the world's record for nine years. The sports world felt that no one would ever run a 4-minute mile. That is why, when Roger Bannister in 1954 posted a 3-minute 59.4-second mile, headlines around the world announced the breakthrough. A quarter century later, his accomplishment was still celebrated, even though he held the record for less than seven weeks. He was the one who had broken the spell.

Let's look at Hägg's and Bannister's performances from a fresh angle, pretending that time is recorded in seconds but not in minutes. Then Hägg's time is 241.4 seconds and Bannister's is 239.4 seconds. No one would have made a big fuss that the 240-second barrier had been broken, for 240 is not a particularly famous number.

But we can compare their records another way. This time, assume instead that a "second" is 1 percent longer than the customary second that we are used to. Then Hägg would have run a 3-minute 59-second mile. He, not Bannister, would have been the first to break the mythical 4-minute barrier. Bannister would be just a footnote in the record books, one of the many runners who once held the record. Such are the quirks of fate and the power of a number.

Numbers ending in a string of zeros seem to cast a special spell. When a car's odometer reaches 49,999.9, everyone watches as all the 9s slowly turn to 0s. Missing this moment is a minor tragedy. Of course, a number ending in 0s is no rarer than a number ending in, say, 37,452, but no one wants to watch 37,451.9 turn into 37,452.

Numbers ending in 0s can actually influence events in the real world. For example, before the last game of the 1941 baseball season, Ted Williams was batting .39955, which rounds off to a beautiful .400. But Williams, unwilling to settle for a fraudulent .400, risking all, insisted on playing the final game. Dramatically, he ended up with .406, the last player to scale that mythic peak of .400.

In 1995, when the stock market reached 4,000 for the first time, brokers felt obliged to explain to their clients, "Remember, 4,000 is just a number."

A string of 0s played a major role in the nuclear arms race. In 1960, when the United States had 68 missiles and the Soviet Union perhaps as few as 4, we decided to build 1,000 missiles. Why 1,000? Did the wisest heads in the Pentagon choose this number after a long strategic deliberation? Not at all.

Instead, one general proposed 10,000 and another 100, but it was felt that Congress would never pay for 10,000, and 100 looked too trivial.

Just what is 1,000? It is ten times ten times ten. Why ten? Because we have ten fingers. That means that if we had only eight fingers, then the generals would have asked for eight times eight times eight; that is, only 512 missiles, a request that would have saved billions of dollars. (If we had only eight fingers, we would use only the digits 0, 1, 2, 3, 4, 5, 6, 7 and write eight as 10 and eight times eight as 100 and never need the words *nine* and *ten*.)

Even the duration of the Gulf War was influenced by a cool number— another string of 0s—as this exchange reported by the *New York Times* shows:

> "Why a cease-fire now?" an irate General Waller asked.
> "One hundred hours has a nice ring," replied General Schwarzkopf.
> Waller uttered an epithet.

Even a complicated-looking number, the very opposite of a number ending in a string of zeros, can cast a spell. When we read that the population of the United States is 256,437,125, our first reaction is amazement that the Census Bureau could achieve such precision. A moment's reflection then convinces us that it couldn't, since, among other things, some ten thousand babies are born every day. As a matter of fact, the Census Bureau doubts that it comes within 2 percent—or 5 million—of the correct figure. But if it reported the population as "somewhere between 251 and 261 millions," its chief would be fired. After all, the number of representatives each state sends to Washington is determined by its population. That's why the Constitution requires a census every ten years. To give the impression of precision and competence, the Census Bureau should avoid numbers ending in lots of zeros.

I've mentioned only a few of the numbers that have shed their cool abstraction, warmed a little, and achieved lives of their own—so rich that they influence our behavior. A few others are 3, 7, 40, and 666, but each reader may find others particularly attractive or alarming. The next chapter shows that just about any number can suddenly become hot and famous, even if for only the proverbial 15 minutes.

··· 3 ···

Hot Numbers

The numbers we see on the front page are not usually cool. They are cited in the midst of major controversies. They are hot numbers. Their effectiveness comes partly from their connotations, for they call to mind the objectivity of science, precision, and logical thinking. These numbers are computed because some multibillion-dollar investment or legislation is the focus of debate. They are a key part of the rhetoric.

If someone in the so-called real world goes to the trouble of calculating a number, it is because that number is of interest: It may influence a decision. No one spends days finding a number just for the fun of it. Instead, people offer a number to help persuade or to win an argument. When the argument is settled, the number cools off. Numbers designed to sway opinion are hot, in contrast to the cool numbers.

A cool number, swept into a heated controversy, can turn hot. For instance, the 1990 census figure for New York City switched from cool to hot when city officials complained that the count was much too low. The undercount (if it occurred) meant that New York would lose millions of dollars of federal funds.

When even a cool number can cast a spell, a hot number can overwhelm us if we are not on guard. So let us take a look at a few of them, the better to protect ourselves from their influence.

A hot number played a central role in getting San Francisco's subway system built. In 1962, the citizens were asked to vote for the largest municipal bond issue in history to pay for the Bay Area Rapid Transit system, known as BART. They were told that by 1975, there would be 258,496 riders daily. That number was critical, for it assured a profit of 13 cents a ride, enough to cover all expenses. It turned out that in 1975 there were only 135,000 riders a day, which meant a loss of $1.31 a ride.

Where, then, did the figure 258,496 come from? It appeared in a report furnished by consultants, so-called transportation experts. Had these experts predicted, "there may be anywhere from 100,000 to 300,000 riders, who can tell?" the bond measure would not have passed. (As it was, it passed by only a slim margin.) That number, 258,496, with its six-figure precision, suggested that some expert had used a formula too complicated for mere mortals to grasp. The precision both reassured and intimidated. How could a number given so precisely not be right?

That number did what it was meant to do: get the subway built. It served its purpose. Whether it turned out to be wrong or right is irrelevant. No one is going to tear up the subway tracks because the number was way off. Only a scholar compelled to browse through old newspapers would bother to look back and check it. The purpose of the number was to win a battle, not to achieve truth. In a way, such a prediction is like a joke, which is a success if we laugh. To ask whether it is true misses the point.

A number that is part of a prediction is especially effective, for two reasons. First, it is hard to discredit a prediction at the time it is made; second, we are in awe of anyone who claims to see into the future.

The basis for a prediction (especially if it was simply plucked from the sky or based on a hunch) should be concealed. The number should be surrounded by such phrases as "the computer model shows" or "sophisticated regression analysis reveals" or "experts have found" or "this is a conservative estimate." Otherwise, exposing the basis for the number offers opponents an avenue of attack, and the number will lose its capacity to charm. As my son, Joshua Stein, advised in an article titled "The Art of Real Estate Negotiations," "the more details the negotiator gives, the more opportunities the Other Side has to find problems. Therefore he should be mysterious when he can."

It is impossible to discuss hot numbers without examining, however briefly, the role of "experts." An expert is supposed to know something that the layperson does not. Consequently, an expert has the license to make assertions without explaining why they are true. Experts in our society play the same role that shamans played in primitive tribes: to suggest that a big gamble, such as a massive program to reduce crime, is a sure thing. Moreover, as the projects of our society grow bigger and stretch further into the future, the more we will have to label certain individuals as experts.

How do we know that a person is an expert? One way is to check the expert's claims. For instance, it's easy to see whether a juggler is an expert.

But how do we tell whether someone is an expert on subway systems? We are forced to decide on the basis of secondary symbols: degrees earned, positions held, manner of dress and speech, quality of posters and overhead slides used. One of the most effective secondary symbols is the use of numbers. That figure, 258,496, not only supported BART, it helped identify the consultant as an expert. It served as the ultimate put-down and conversation-stopper.

One species of expert, the meteorologist, often accompanies a prediction with a percentage, as in, "there is a 30 percent chance of rain." (Studies show that after about 30 percent of such predictions, it does rain.) It makes sense that anyone offering a prediction should include a percentage. If so, the prediction about BART ridership might have read, "There is a 50 percent chance that in 1975 daily ridership will clear 250,000." In this form, the prediction is as cool as a weather forecast. It no longer presents a hot, persuasive number. No one would vote for a billion-dollar bond on the strength of such a wishy-washy assertion.

During the Vietnam War, one reason stated for continuing it was that a North Vietnamese victory would result in a bloodbath, with 500,000 victims. After the war there was a large-scale "reeducation" program but no bloodbath. Where, then, did that number come from? It turns out that during the war, the Vietcong executed five leaders in a small village. Someone, knowing that the population of South Vietnam was 100,000 times that of the village, had simply multiplied 5 by 100,000. That often quoted number, 500,000, did its part in prolonging the war. The prediction, "There would be a big bloodbath," would have had little effect. Adding the number gave it a punch, illustrating the maxim that one number is worth a thousand words.

The number 500,000, detached from its origins, took on a life of its own like a bird leaving its nest, and it turned into a Fact. However, it is the nature of a hot number that its source be shrouded in mystery, at least until it serves its purpose.

President Clinton claimed a reduction in the 1994 federal deficit of $102 billion. Though that number sounds easy to compute, even it is controversial. (Behind a hot number is usually an interested party, trying to push it up or down. Hot numbers tend to be as elastic as a rubber band.) The deficit in 1994 was $203 billion. You would expect to subtract that from the 1993 deficit to find the reduction. That is not how it is done. Instead, you subtract $203 billion from what the deficit in 1994 would have been if the president's program had not been enacted.

The White House estimated that hypothetical number as $305 billion and hence arrived at a reduction of 305 − 203 = $102 billion. The Congressional Budget Office, on the other hand, used an estimate of only $286 billion for the hypothetical budget and concluded that the reduction was only 286 − 203 = $83 billion. It even had the audacity to give the president credit for only $33 billion, crediting a booming economy for $50 billion. However, since presidents traditionally take the credit (or the blame) for whatever occurs on their watch, President Clinton should claim the $50 billion as well.

A similar massaging of a hot number is possible when an automaker announces, "This year's model costs only 2 percent more than last year's, an increase less than the rate of inflation." Yet a comparison of actual prices shows an 8 percent rise. How can this be? It turns out that features that were optional the previous year are standard in the new model. The comparison is made between comparably equipped cars. The 2 percent figure contrasts the new model with one that never was—a hypothetical model—just as the federal deficit was compared with a hypothetical deficit.

Even the field of education heats up many a cool number. For instance, in the United States there are 180 school days per year, while in Japan there are 220 days. So, in order to improve the performance of our students, some educators recommend extending the school year. But things are not so simple. In the United States, there are 1,003 instructional hours; in Japan, only 875. However, there is more to it. The National Education Commission on Time and Learning found that students in the United States spent only 41 percent of the time on core subjects, such as mathematics, English, science, and history. By the time they graduate, U.S. students will have studied these subjects for 1,460 hours, whereas Japanese students will have spent 3,170 hours on core subjects. Each comparison supports a different reform: lengthen the school year, shorten the school day, or devote more time to the core. The desired conclusion controls the statistic presented. Perhaps none of the numbers tells the real story. Instead, the time spent on homework may be the key—or even something that cannot be measured, such as the parents' attitudes.

Newspapers often publish charts that rank the states by their expenditures for schools, as measured in dollars spent per pupil. I would think that this would involve only cool numbers: Just divide total expenditure by the number of pupils. Even these numbers, however, turn out to be hot, and they can be raised or lowered depending on whether the interested party is arguing for more school money or lower taxes.

First, there are no uniform definitions of *expense* or *pupil*. For instance, California does not include as an expense the funds generated by its lottery for use in education. However, it counts an excused absence as "present," which alone has a 6 percent effect on the budget. Some other states, wanting to rank high on the chart, include as an expense the cost of a road built to provide access to a school. To further complicate matters, the cost of living varies from state to state. Once again, if you don't see the nitty-gritty—all the steps in the calculation—you can't be sure what the numbers mean.

The *Wall Street Journal* of June 22, 1993, contained a table of numbers titled "Money Doesn't Help." This table listed for each state "Average per pupil expenditure" and "SAT rank." The ten states that spent the most ranked below the 11 states that spent the least. This sounds like strong evidence for an open-and-shut case of reducing school funding in order to improve education. That conclusion would appeal to every taxpayer.

Unfortunately, there is more than meets the eye. Howard Wainer, in an article called "Does Spending Money on Education Help?" looked beneath the numbers to another layer of numbers that didn't appear in the table. He examined the percentage of students who took the SAT examination. For instance, in Iowa, a low-spending state, which had the highest SAT rank, only 3 percent of the students took the examination. In Connecticut, a high-spending state that ranked very low, 78 percent took it. As Wainer points out, "If we chose the top 3% from Connecticut SAT-takers, their average scores would smother the currently top-ranked Hawkeyes."

Clearly, we cannot be too cautious in accepting a number at face value. Yet with so many controversies swirling around us, it is hard to probe beneath their surface. In the midst of a hot issue a hard, unassailable fact is as rare as diamonds. Even if clothed in the protective armor of numbers, it is open to question: How was the number found? How were the terms defined? What is the raw data? If there was a poll, how were the questions phrased? Was the number also computed by an independent researcher, as is the case in the hard sciences?

A men's support group, wanting to point out that women are not the only ones who have it bad, announced, "More than half of the women on death row had murdered their husbands, but only a third of the men there had murdered their wives." This was correct: four of the seven women, but a third of some 2,400 men. This shows that a fraction is a good place to conceal numbers that would otherwise be unimpressive.

Sometimes it is hard to realize that a number that seems cool is hot. This is particularly true of numbers intended to reassure us. When we hear the results of a poll and are told, "The margin of error is plus or minus 3 percent," we feel we know how accurate the poll is. But as Humphrey Taylor, chairman of Louis Harris polling, advises, "This is not true. The history of polling is replete with 'errors' that exceed the so-called margin of error." Even so, news anchors will continue to "guarantee" that the possible error is under control. Doing so gives the poll an aura of mathematical precision.

One of the hottest numbers is the estimate made by the Centers for Disease Control (CDC) of the number of HIV cases. Because of the issue of confidentiality, the data are shaky, and estimates vary from 600,000 to 1.2 million. Advocates of increased funding to fight AIDS want CDC to raise its official estimate. Those who have introduced programs to reduce the spread of AIDS want the estimate lowered to demonstrate the success of their programs. Furthermore, as one specialist observed, "So many people have so much invested in the epidemic that anything that makes it look like it is going away causes panic among the investors." The CDC is in a no-win dilemma, bound to offend some powerful group.

To convince yourself that a particular hot number can be maneuvered up or down, just imagine calculating it yourself. It won't take long to see how vulnerable and elastic it is. Though a number may at first appear overwhelmingly convincing, it may be the weak point in an argument. The best defense against a hot number is to insist on seeing every single gory detail in the computations that produced it: the definitions, the questions in the surveys, the assumptions, the data. Of course, once those details are exposed to the light, the hot number cools, sheds its persuasive power, and may no longer be of use to anyone.

··· 4 ···

Don't Do
a Number on Me

G iving something a name does not assure us that it exists. Scientists once thought that fire was made of a substance they called *phlogiston*. They also thought that light traveled in a medium, just as sound travels in air. They called that medium the *ether*. Eventually, like children who have learned there is no Santa Claus, scientists found out that there is no phlogiston and abandoned the notion of an ether. Yet we feel that if there is a word, there must be something named by the word. That's the "if there's smoke, there's fire" point of view.

But there's another quirk of our minds that I want to explore. We tend to describe things by numbers and preferably measure them by a single number, whether it's the state of the economy or the intelligence of a person. Unfortunately, this habit can lead to nonsense and even harm. Much as I love numbers, I am going to argue that they are overused and suggest a way to spot a common form of their abuse.

Pretend for the moment that we have never heard of numbers. Even so, we can tell when one person is taller than another. Just stand the two people back to back and look at them. Now let numbers enter the scene, and with the aid of a measuring tape, measure the height of each person. Then the person with the bigger number is taller. This method enables us to decide which person is taller, even if they are far apart, even in different cities. *Height* is a fairly simple notion and can be described by a *single* number. For that reason we will call height *one-dimensional*.

The weight of a person is also one-dimensional. Without using numbers, we can tell when one person weighs more than another. Put the two people at equal distances from the axis of a balance and see who goes up and who goes down. It is natural to expect there to be a method of assigning numbers to weights.

Next consider all possible colors. Can each color be described by a single number? In other words, can all colors be arranged along a line? To answer this, recall how artists combine the three primary colors—yellow, blue, and red—to obtain other colors. For instance, a combination of 50 percent yellow, 50 percent blue, and 0 percent red gives green. Or a combination of 60 percent yellow, 40 percent blue, and 0 percent red gives a somewhat lighter green. Mix 50 percent yellow, 0 percent blue, and 50 percent red, and you get orange. Or you might take 20 percent yellow, 30 percent blue, and 50 percent red to obtain yet another color. Each choice of the three percentages that add up to 100 percent yields a different color. Actually, the choice of just two percentages describes the color, since it determines the percentage of the third color. For instance, if you have 18 percent yellow and 32 percent blue, then you must have 50 percent red.

Unlike height and weight, color cannot be described by a single number. It is not one-dimensional. It is impossible to arrange all colors on a line. This makes some trouble for paint manufacturers, who display their sample color chips spread out like squares in a checkerboard.

It also makes trouble for art students, who can't afford dozens of tubes of different colors. To cut the expense and clutter, they may buy just a few tubes and make other colors by mixing those few. But if they return to work on their painting a week later, they may not remember the exact combination that produced a certain color. No wonder that entire courses in art school are devoted to the theory and practice of color.

But color is far more complicated, and cannot even be described by two numbers. Artists must pay attention also to "value" and "chroma." Value describes how dark or light a given color is, while chroma refers to its intensity. Some Renaissance artists, working with a limited array of colors, emphasized value. For instance, Da Vinci made a portrait almost exclusively in one color, but achieved dramatic effects by strong contrasts of light and dark values.

Instead of height, weight, and color, consider strawberries. Food scientists have discovered over 1,000 chemicals in a strawberry of which only 15 significantly influence the flavor. Flavor is a combination of both aroma and taste. (When a cold plugs up your nose, flavors may even disappear.) Food scientists and physiologists, in spite of much study, still do not understand the mechanisms by which a person senses flavor. Perhaps many numbers, even far more than 15, may be needed to describe a flavor sensation, or perhaps flavor will never be described by any collection of numbers.

With heights, weights, colors, and flavors behind us, let's take a look at the collection of possible "intelligences" of people. Whatever "intelligence" might be, it is allegedly measured by a single number, the so-called Intelligence Quotient, or IQ for short. If intelligence really can be measured by a single number, then it must be as simple an attribute of a person as height or weight—just another one-dimensional quantity. That would tell us that the mind of a person is much simpler than a color or the flavor of a strawberry.

But what does the word *intelligence* mean? According to one dictionary, it is the capacity to acquire and apply knowledge. I would put it differently, as the capacity to meet the challenges of life. That involves judgment, experience, perseverance, emotional stability, communication skills, reading ability, the capacity to apply acquired knowledge, and so on. The word *intelligence* may, like phlogiston or the ether, describe such a vague concept that we could just as well assume that it describes nothing at all. That means that an IQ test may measure only the ability to take an IQ test. (People who perform at a level inconsistent with their IQ scores are called *over-achievers* or *under-achievers*. This trick manages to transfer the blame from the test to the person who exhibits such perverse and peculiar behavior.)

The psychologist J. P. Guilford, using a statistical analysis, concluded, "There are at least fifty ways of being intelligent. Simplicity certainly has its appeal. But human nature is exceedingly complex and we may as well face the fact." That means that at least 50 numbers are needed to describe intelligence (if it can be described by numbers at all).

It should not come as a surprise that a study that followed 379 Boston schoolchildren from 1951 to 1992 found that "parents' putting pressure on the children to achieve was a stronger predictor of earnings and job achievement than were childhood IQ scores." Another long-term study concluded that character traits, such as "will power, perseverance, and desire to excel," play a key role in predicting success.

I suspect that if IQ testing were abandoned, civilization would not notice the loss. After all, we produced the Renaissance, the Constitution, and the industrial revolution and developed the train, automobile, and airplane before the first IQ test was offered. (It must have been an IQ test that put me in leatherworking in the seventh grade, where I made a purple wallet. After several weeks, my mother complained to the authorities, and I was switched to Latin.)

In the nineteenth century, psychologists claimed that they could measure intelligence by the volume of the brain. This approach, which went under the impressive name *craniometry,* has vanished into the dustbin of history. Perhaps the IQ, a fruit of the twentieth century, will suffer a similar fate. For an amusing and thorough telling of the various efforts to reduce intelligence to a number, read Stephen Gould's *The Mismeasure of Man.*

Before we dare to describe any property by a single number, we should first ask ourselves, "Is it really so simple that it's just one-dimensional, like a person's height?" The key question that then faces us is, "Is there a simple way to tell whether one object has more of that quality than another object?" In the case of height and weight, we saw that the answer was yes.

Does it make any sense to ask, "Who is more intelligent, Mozart or Einstein?" That the question is ridiculous warns us not to expect intelligence to be measured by a single number. It may not even be described by any group of numbers.

Matters grow more absurd when the IQs of various races are compared. Molecular geneticists point out that there has been very little time to achieve much genetic difference between races, even though they may have been separated by 100,000 years, which is just too brief an interval for the brain to undergo significant genetic change. There is far more variation among individuals within a race than between races. If assigning an IQ to an individual is already a shaky enterprise, assigning an average IQ to a race will certainly produce nonsense.

Numbers charm and intimidate. In our society, the user of numbers is one up on the user of mere words. A leading political scientist once presented a formula that produced a single number for allegedly measuring the "political instability" of a country. Oddly, according to the formula, France was unstable and South Africa was stable, at a time when anyone who read a newspaper knew the reverse was true. Not until the political scientist was nominated to the National Academy of Sciences was the formula analyzed (by a mathematician) and exposed as absurd. The political scientist's main defense (which failed) seemed to consist of attacks on the character of the mathematician.

Following is another example showing that a single number can seldom reveal the essence of anything complex.

In a chart titled "A Sampling of American Paychecks," the *New York Times* compared the incomes of 49 occupations for the year 1994. In order to make the comparisons, it had to describe each income by a single number. It chose "median weekly income" as the measure. This number was $996 for physicians, which is about $52,000 a year, a figure that struck me as quite low.

I wrote to the Bureau of Labor Statistics, which had supplied the data for the chart. In the printout they sent me was the information that the "mean" weekly income for their sample of physicians was $1,586. (According to the American Medical Association, the mean in 1991, when incomes were about the same, was $3,281.) Now, *median* and *mean* are quite different concepts. A median of $996 tells us that half the physicians earned less than that figure and half earned more. The mean is the average of the incomes, that is, the sum of all the physicians' incomes divided by the number of physicians. That the mean is so much larger than the median in this case is due to the presence of a sizable number of doctors with large incomes.

The data, based on a sample of 358,000 full-time salaried physicians, break down as shown in the graph below.

These numbers tell much more than any one number can. Any attempt to describe the earnings of an occupation by just one number is bound to sacrifice essential information. Even the numbers in the table do not tell the full picture. For instance, heart surgeons earn three times as much as family physicians. It is not easy to reach the bedrock of reality, no matter how many numbers we use. A hard fact is a rare jewel, whether told in words or numbers.

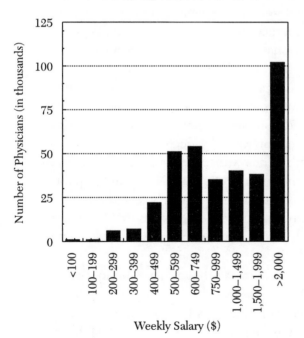

The impulse to reduce a concept to a single number should be resisted. Certainly it would be pleasant to turn the vague into the precise, the complex into the simple, and to make real what may be only a figment of our imagination. Unfortunately, what is important in life is seldom one-dimensional.

··· 5 ···

Anecdote versus Number

When the Surgeon General announces, on the basis of many massive studies stretching over decades, that smoking kills 1,000 Americans a day, we may respond, "Uncle Joe smoked two packs a day and lived to the ripe old age of 101." With this reply we may feel that we have debunked the Surgeon General's claim. After all, our personal experience, though it is just a small sample, looms large, vivid, and convincing.

Where do we get our opinions? Often they are generalizations based on a few cases. Seeing a neighbor do this or that, we may generalize to all redheads or all skinny people. In this approach, we trust the evidence of our own senses. Or we may put our trust in a leader who generalizes for us, citing an instance that may be real or imagined. In either case, we draw conclusions from limited information. Call this method *reasoning by anecdote,* which tends to be based on examples chosen for their drama, whether worst-case or best-case scenarios. It is surprising how many debates on public policy are conducted in terms of a couple of extreme cases, real or imagined. By its very nature, this approach will usually have a built-in bias.

In a debate on the welfare system, one person may cite the case of a recipient who defrauded the government. Another, in rebuttal, may mention the middle-class woman who, in one week, lost her husband and her job and needed welfare to survive.

In contrast to reasoning by anecdote, there is the scientific method, designed to avoid drawing conclusions from too few cases. Numbers, rather than words, are its main tools. As statisticians put it, "In God we trust. All others supply data." Instead of drawing a conclusion from extreme cases, one pays attention in this approach to the vast majority of cases. In a sense, this is the very opposite of the anecdotal method, for it will sometimes delete the extreme cases as misleading "outliers." A few examples will contrast the two approaches.

Do psychics really help police solve crimes? One way to answer the question is to think of some famous crime where the police consulted psychics and recall whether the psychics' insights were helpful. The answer depends on which case happens to come to mind. This is the usual approach in daily life, in which debaters fling anecdotes at each other. The Los Angeles Police Department, wanting to settle the issue, decided to use the scientific approach.

This is how police psychologists N. Klyver and M. Reiser conducted their experiment. A detective selected six murder cases and chose evidence from each crime scene. Twelve psychics, 12 detectives, and 11 college students were instructed to write a description of the victim and murderer on the basis of the evidence.

The responses of the three groups differed markedly in quantity, with the psychics' averaging a page and a half singlespaced and the others' only a quarter of a page. The psychics' responses also differed in character, for they presented their intuitions with greater detail and confidence. Typical of their visions:

> I see some kind of hospital, operation room . . . wounds from three bullets
> . . . I see this clearly.

or

> Last name of Price. A problem with his right wrist. I keep getting the
> number 62.

or

> August 9th. Something significant about August 9th . . . Plays tennis.

Since the psychics' reports were much longer than those of the "control" groups—the detectives and the students—chance alone should give them more hits. Despite this advantage, the psychics' information was no better. No one—psychic or control—produced any useful information, such as names, license plate numbers, or locations.

The number of accurate statements made was as follows: psychics, 34; detectives, 27; students, 39. The differences are not statistically significant.

Klyver and Reiser concluded: "The use of psychics is unlikely to produce useful information. Perhaps the compelling manner in which self-identified psychics tend to present their information may account for some of the positive beliefs about psychic abilities in law enforcement. Individuals may be persuaded more by the dramatic character of the information than by its merit."

We don't need to be statisticians to conduct and interpret such an experiment. The numbers tell the story.

In a similar spirit, the logician Martin Davis has proposed a way to test the skill of an astrologer. Usually an astrologer is told a person's zodiac sign and deduces things about that person's character and life. But turn this around. Have an astrologer figure out your sign from information about your character and life. (The astrologer can ask any questions whose answers do not give a clue about your birthday.) Of course, since there are 12 signs, the astrologer has 1 chance in 12 of being right just by guessing. To reduce the role of chance, have, say, 24 people ask for their signs. Even lacking astrological gifts, a person would guess right about two times. However, the chance of getting at least 12 out of 24 right is quite small. Davis has not performed the experiment, but it would not be difficult to carry out, perhaps at a party. The key ingredient is a confident astrologer.

Before we leave the realm of the paranormal, I will describe a simple experiment I once did, which I invite you to repeat.

Around the first of the year, the tabloids you see as you wait in the grocery checkout counter offer many predictions by psychics for the new year. I bought a copy and kept it. At the end of the year, I checked how many predictions turned out right and how many were wrong. (I didn't count the ones that were so ambiguous that I couldn't decide.) The batting average was a mere 5 percent.

There was no point in telling the tabloid, for I knew that its readers enjoy the predictions: the more astonishing the better. But you can carry out a similar check on any of the seers who participate in the annual prediction ritual, whether psychics, economists, or political pundits. As you check them a year later, you will feel that you are violating an unwritten code that says it is rude to look back.

When I watch the local evening news, the first five or ten minutes often consist only of crime scenes cordoned off with yellow tape. Even if a murder at a convenience store takes place 2000 miles away, I will be shown the gory details. So, during the commercial break I check that my doors are locked. But is there a dramatic rise in murders or just a jump in the time television devotes to it? This question, too, can be examined with the aid of numbers.

We don't have to wait for the information superhighway to find the numbers. We are already in a superswamp of information. The problem is not a lack of facts but a lack of time to digest them. As an example, the Federal Bureau of Investigation (FBI) annually publishes the *Uniform Crime Report,* which contains more numbers than anyone could absorb. Let's see what it tells us about the rise or fall in the number of murders in the United States over a period of 20 years.

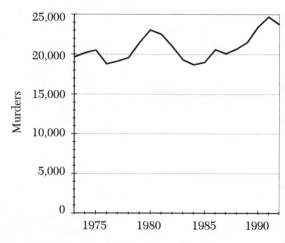

The graph at right shows the number of murders each year from 1973 through 1992. There was a low of 18,700 in 1984 and a high of 24,700 in 1991. There is a slight upward drift. However, during those 20 years, the population grew from 210 million to 255 million. Conveniently for us, the FBI took this increase in population into consideration and provided another table, showing the number of murders per 100,000 inhabitants. The rate started at 9.4 in 1973 and ended at 9.3 in 1992. The following chart shows the rate during the same 20 years.

This picture sends a different message than the other one. Instead of an upward trend, it portrays a murder rate going up and down around

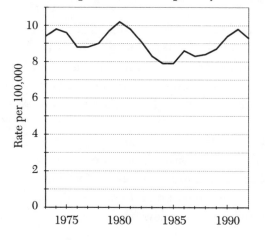

the rate of 9 per 100,000. It certainly does not show a sudden rise. Instead, there had simply been a rise in the time that television news devoted to crime. Moreover, in 1995, the homicide rate in several large cities dropped dramatically. This shows how hard it is to get a fact straight and how easily we can be led into error.

The reader can browse through the data on "Aggravated Assaults,"

"Burglaries," "Motor Vehicle Thefts," and so on. Some of the rates have risen, some have fallen. There's enough to buttress any case you want to make.

Before I left the FBI reports, I came upon "Justifiable Homicides," which is the killing by a private citizen of a felon during the committing of a crime. In 1990, 276 of these involved firearms. These homicides include the case where a homeowner shoots a burglar. Then I looked at the *Statistical Abstract of the United States,* another wonderful storehouse of all sorts of numbers, to find that there were 1,416 accidental deaths from firearms that year. According to an annual publication of the National Safety Council, some 800 of these were in the home. (Clearly, we are already knee-deep in a glut of numbers.) This means that for every burglar killed with a firearm in the home, about three innocent people lose their lives. That ratio raises a serious question about the benefit and risk of keeping guns around the home.

Is driving at night more dangerous than by day? I had an opinion based on anecdotal evidence, but I thought that the scientific method would apply. I decided to compare the number of accidents each hour with the corresponding amount of traffic. For instance, if a particular hour of the day had, say, 6 percent of the day's accidents but only 3 percent of the day's traffic, then that would be a high-risk hour. On the other hand, if an hour had 3 percent of the day's accidents and 6 percent of the day's traffic, that would be a low-risk hour.

As it happens, the California Transportation Department automatically records the amount of traffic hour by hour on many roads and highways. The Highway Patrol keeps a record of fatal vehicle accidents, also hour by hour. Wanting to compare them, I requested several pounds of their computer printout. The following graph summarizes their data.

It shows the *percentage per hour.* For instance, if the amount of traffic were the same throughout the 24 hours of the day, then each hour would have ¹⁄₂₄, or about 4 percent, of the whole day's traffic. However, the traffic varies. At night it is light. By day it is heavier, with the morning and evening rushes showing up as two peaks.

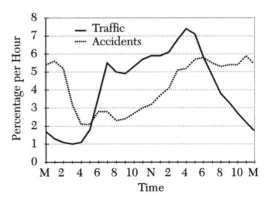

The chart of accidents does not resemble the one for traffic. From 6:00 A.M. to 6:00 P.M., it looks like the traffic chart, which is to be expected. But why is it so high at night, when there is so little traffic? It could not be due just to the darkness, since the accident rate drops a good deal from 2:00 A.M. to 4:00 A.M., at a time when the traffic declines only a little.

To help resolve the mystery, I defined the *risk*, or *danger*, at any hour of day as the number you get when you divide the accident percentage by the traffic percentage. For instance, if an hour had 6 percent of the accidents but only 3 percent of the traffic, the risk would be ⁶⁄₃, or 2. On the other hand, if an hour had only 3 percent of the accidents but 6 percent of the traffic, then the risk would be only ³⁄₆, or 0.5. During an hour when there were lots of accidents and only sparse traffic, the risk would be large, greater than 1. On the other hand, if there were few accidents and dense traffic, the quotient would be small, less than 1. The following graph shows how the risk varied during the 24 hours of the day. It rises gradually from

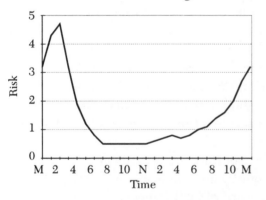

6:00 A.M. to 6:00 P.M. and then more steeply from 6:00 P.M. to 2:00 A.M. Why?

As might be expected, I then looked at the amount of alcohol in the drivers involved in fatal single-car accidents. Even this information is stored in computers.

A blood alcohol content (BAC) of at least .10 percent (1 part alcohol in 1,000 parts blood) means that the driver is intoxicated—"drunk," for short. (California has since reduced the threshold to .08 percent.) A BAC of 0 we call "sober." A BAC in between we call "drinking." The next two graphs show how the numbers of drunk and sober drivers varied during the day. I use a different scale in the second graph, since there were so many fewer drunk drivers.

What I found fascinating is that the drunk graph has practically the same shape as the risk graph. That is no coincidence. In the period

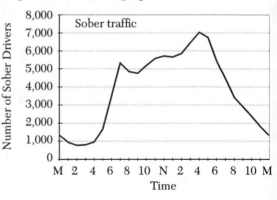

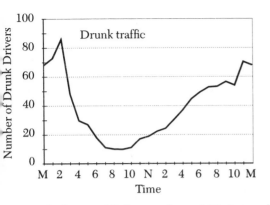

when the risk reaches its peak, some 80 percent of the fatal accidents involve alcohol. During the day, this figure drops to 20 percent.

I couldn't resist estimating the risk factor for drunk drivers. Combining the printouts with other data, I found that a drunk driver at any time of day or night has a risk factor about 100 times that of a sober driver.

When I published some of these conclusions in the *Sacramento Bee,* the Associated Press picked up the story. For a couple of weeks, newspaper reporters and talk show hosts interviewed me by telephone. While everyone knew that drinking and driving don't mix well, it came as a surprise that the mix was so dangerous.

Numbers, combined with a little arithmetic, can replace anecdotes and rhetoric in analyzing many other issues; for instance, "Is professional boxing a ticket out of poverty?" For the champions it may be, but what about the less conspicuous boxers who end up punch-drunk and without any usable skills? Their stories do not appear in *Sports Illustrated.* To get the whole picture, shouldn't we balance the number of boxers who escape poverty with the number who, failing to develop an employable skill, become more deeply mired in poverty? (After all, no occupation other than boxing requires workers with the talent for knocking people unconscious.) The same question can be asked of basketball, where fewer than 1 percent of the half million young men who play high school basketball win scholarships to college. Of those who play in college, only a few make it to the professional leagues. The sports enthusiast might want to find the numbers needed to answer these questions.

It is surprising how many issues that we usually address in words can and should be examined by the scientific method, with the introduction of numbers. The reader will find a host of examples among the controversies raging in today's newspaper—controversies usually settled by such rhetorical tricks as name-calling, inventing hypothetical examples, or citing extreme but atypical cases.

The sheer size and complexity of the issues that face our nation ought to demand an appropriate updating of the way we settle them. We need more signs of such a modernization.

It Ain't
Necessarily So

Mathematics gathers more than its fair share of erroneous myths. This may be due to a combination of two circumstances. On the one hand, everyone comes in contact with a bit of mathematics almost daily; on the other, most of mathematics is remote from the everyday world, situated in a realm of special symbols and arcane definitions. The less we know about a subject, the less encumbered we are by the facts; we therefore remain freer to form opinions about it. Moreover, the shakier the foundation for our beliefs, the more strongly we often hold and defend them, as though the strength of our attachment is evidence for their truth.

This chapter debunks some of these myths and raises doubts about others, charming and common though they may be.

There is a gene for mathematical talent.

This belief, entertained mostly in the United States, is far from universal. I have never seen any evidence to support it. Instead I have a much simpler explanation for why one person does better than another in learning mathematics: efficient *time on task,* that is, the amount of time spent in well-organized study. Any normal person who turns off the television, attends to details, and is persistent can learn mathematics. I am talking about learning mathematics that already exists, not about creating new mathematics. I base this conclusion on my experience teaching several thousand college students and tutoring in school. A person's height, weight, hair color, race, gender, or national origin doesn't matter. This observation holds for kindergartners up through Ph.D. candidates and beyond.

Of course, there are some cases where the children of mathematicians become mathematicians. But there are also circus acrobats whose children become circus acrobats. No one proposes that there is a gene for circus acrobatics.

There is nothing new in mathematics. It's a dead subject.

Well might people believe that all of mathematics was invented and polished centuries ago, for nothing in the school curriculum suggests otherwise. The decimal system of notation for the whole numbers, which goes back to Hindu mathematicians of the year 600, was introduced to Europe by Leonardo Fibonacci in 1202. Decimal notation for numbers between 0 and 1 goes back, in Europe, to about 1600, but it was known to Arab mathematicians in around 1000.

Just about all of high school geometry was collected in one book by Euclid around 300 B.C. Trigonometry was well developed by Greek astronomers some 1,800 years ago.

Negative numbers and the use of letters to stand for numbers go back to the seventeenth century. Even the practice of using the front of the alphabet to denote constants and x, y, and z to denote varying or unknown numbers dates back that far. Almost all the mathematics a pupil sees, from kindergarten through calculus, consists of antiques at least three centuries old. That doesn't mean that they are obsolete. Far from it. They still provide the front door to the temple of mathematics. There is nothing wrong in teaching or learning mathematics that is, say, 600 years old. That is still the mathematics that most people need. After all, there is no stigma attached to uttering such useful words as *item, exit, agenda,* or *credit,* straight out of Latin, all of which have been around for some 2,000 years.

But mathematics, rather than resting on its laurels, has continued to grow. In fact, it is still in its golden age, which stretches without interruption back to the Renaissance. If I were to pick a starting date for this period, it would be the invention of the printing press by Johannes Gutenberg in 1439. The shift from handwritten manuscripts to printed books suddenly made ancient knowledge and modern discoveries available to a broad public.

To give some idea of the vitality of mathematical research, I will use a few numbers. Every month the *Mathematical Reviews* requires some 600 pages just to present brief summaries of the new pure and applied mathematics. In the course of a year, it refers to the work of some 50,000 mathematicians. Though most of this research will not become a permanent part of mathematics, a significant portion will endure. It is impossible to predict for most of the new work which will last and which will vanish from view. Sometimes, though, when an insight solves a famous problem or recasts the foundation of a major field or finds a valuable application outside of mathematics, it is clear that it will not be soon forgotten. Suddenly, the earlier results on which the discovery depends, which may have seemed of little interest, turn out to have lasting significance.

All that mathematicians do is study numbers.

Many mathematicians, such as logicians, topologists, algebraists, and theoretical computer scientists, may never do any calculations, may never even bump into a number. Some of the most important discoveries do not involve numbers. For instance, take L. E. J. Brouwer's (1881–1966) "fixed-point theorem" of 1926, which I will state informally.

Imagine that you have a circular piece of paper lying on a table. Pick it up, fold and crumple it as often as you wish, but don't rip it. Then put it, still folded up, back within the circular place on the table where it was at first. Brouwer says that at least one point in the paper is then back where it was at the beginning or lies directly above its starting position. This is far from obvious; the proof is part of the field called *topology*.

Here is another example, which arose in game theory and in physics. Say that you are allowed to form "strings" of letters (nonsense words) using the letters *a, b,* and *c.* For instance, *accbacbac* is a string. This particular string happens to contain a shorter string that consists of two copies of a string. For instance, it contains the string *cc,* which is the string *c* repeated. It also contains the string *cbacba,* which is *cba* repeated. It also contains *bacbac,* which is *bac* repeated. However, the string *abacbabcab* has no repeated string. The question is, "How long can a string be if it has no repeated string?" Surprisingly, there is an endless string that contains no such repetitions. Note, again, that numbers play no role in this question. (Here is an amusing puzzle in this vein: If you are allowed to use only the letters *a* and *b,* how long a string could you make that has no repeated substring?)

Mathematicians spend all day at their computers.

Many mathematicians don't even own a computer, while many use one only as a word processor. But some do indeed use a computer to carry out computations that would be onerous or impossible to do by hand in a lifetime. Computers have provided information that has led to new insights and questions. In this respect, they serve mathematicians in the same way that telescopes serve astronomers. I'm sure that astronomers do not spend all day or night looking through their telescopes.

Mathematicians start with axioms, see what theorems follow, then look for examples.

That cold, orderly way is not how new mathematics comes to be. If it were, then mathematics would soon dry up.

Here is how new mathematics is usually created. It begins with someone trying to explain or describe some natural phenomenon or a similarity between two mathematical ideas, or with someone trying to make sense of several examples or a mass of calculations. The examples come first, not last.

Then one looks for some assumptions that could, with a bit of reasoning, explain the riddle. At this point there is a theorem, which is made up of an assumption and a conclusion, together with a proof. Next, one tries to whittle down the assumptions to a bare minimum, for the sake of elegance and for a deeper understanding of the discovery.

After proving several related theorems, one tries to build a coherent structure, which consists of a few assumptions and some precise definitions, from which all the theorems follow. These assumptions are called axioms. All told, then, the actual order is examples, theorems, axioms—the very reverse of the myth. By the way, the first person to collect theorems within an axiomatic system was Euclid, around the year 300 B.C.

Complicating matters is that mathematicians may follow a path that leads to a dead end. For instance, they may think that a problem is, say, geometric, when it turns out to be algebraic. This is the case with the ancient Greeks' problem of trying to construct a 20° angle with a straightedge and a compass (for drawing circles).

Using only those two tools, they easily constructed a 90° angle and then by repeated bisections obtained 45° and 22½°. By constructing an equilateral triangle, they could make an angle of 60° and then, by bisection, 30° and 15°. But they could never get 20°, hard as they tried. That means they couldn't trisect a 60° angle, nor could they build a regular nine-sided polygon (since they couldn't get an angle of $^{360}\!/_9 = 40°$). These problems all sound geometric, but their solution turned out to be algebraic.

P. L. Wantzel (1814–1848), in 1837, showed that no one will ever be able to construct a 20° angle with those two tools, because a certain number (the cosine of 20°) cannot be obtained from the whole numbers by the four operations of arithmetic together with the taking of square roots. His argument is completely algebraic.

So a mathematician, like a detective, a mountain climber, or an explorer, must stay nimble, ready to take a fresh route when one approach fizzles out.

Mathematicians are over the hill by age 30.

It is often said that mathematicians do their best work before reaching the age of 30; after that, it's downhill all the way. This myth probably grew out of the lives of two great mathematicians, the Norwegian Niels Abel (1802–1829) and the Frenchman Evariste Galois (1811–1832). In spite of the brevity of their lives, both made major contributions to mathematics, with Galois essentially founding modern algebra. Partly because of these two mathematicians, G. H. Hardy (1877–1947), in *A Mathematician's Apology,* commented, "No mathematician should ever allow himself to forget that mathematics, more than any other art or science, is a young man's game."

Mathematics, like music, chess, tennis, figure skating, swimming, or gymnastics, can be done very well by teenagers. I am no more impressed by a 15-year-old mathematics prodigy than I am by a 14-year old tennis pro or by a 16-year-old Olympic gymnast. Nor am I astonished that Bizet composed his Symphony in C Major at the age of 17 or that Mendelssohn composed the Overture to *A Midsummer Night's Dream* at the age of 18. While these accomplishments demand great talent, devotion, and hard work, they do not require a broad understanding of the human condition. However, if a teenager were to write a profound play or novel, with deep character development, then I would be astonished, for creating such a work requires insights into the human mind, insights that come only with many years of living.

Whether a child gets such a head start is usually a question of circumstance. For instance, when Abel was 15, his harsh schoolmaster was replaced by an enlightened mathematician who guided Abel's reading. Without that switch, I doubt that we would ever have heard of Abel. The mathematician Joseph Bertrand (1822–1900) entered the prestigious university, the École Polytechnique, at the age of 11. That sounds impressive until you learn that he was guided by his brother-in-law, the mathematician, J. C. M. Duhamel (1797–1872).

Also, the ability to create mathematics does not inevitably fade away after the age of 30, as a few examples will show.

Andrew Wiles settled the 300-year old Fermat's theorem, the most famous problem in mathematics, in 1994 at the age of 41 after working on it for eight years. (He proved that if n is a whole number, at least 3, then there are no whole numbers, x, y, and z, such that $x^n + y^n = z^n$.)

Wolfgang Haken was 48 and Kenneth Appel 44 when, in 1976, they settled the four-color problem, which dated back to 1852. (They proved that any map drawn on a piece of paper can be colored using at most four colors in such a way that any pair of countries that share an edge along their borders have different colors.)

Louis de Branges was 52 when, in 1983, he confirmed the Bieberbach conjecture in complex analysis, which was posed in 1916. Many mathematicians had come up with wrong proofs, and the saying became, "The Bieberbach conjecture isn't difficult; I have proved it dozens of times."

Roger Apéry (1916–1994) was over 60 when he found, in 1977, what has been called a "miraculous and magnificent" proof that the sum of the reciprocals of all the cubes is not a fraction. That is, if you keep on adding up the numbers $1/1^3$, $1/2^3$, $1/3^3$, $1/4^3$, . . . , the sums will get closer and closer to a certain number, and that number cannot be written in the form a/b, where a and b are whole numbers. It had been known for over a century that the sum of the reciprocals of all the squares is not a fraction. In fact, this sum is $\pi^2/6$, a discovery Leonhard Euler (pronounced "Oiler") (1707–1783) made in 1743.

Many mathematicians continue to do first-rate work into their seventies or even later. I. M. Gelfand, a Russian mathematician who migrated to Rutgers University, continues to make significant contributions in his eighties. In this he resembles Giuseppe Verdi, who wrote *Falstaff*, considered to be one of his finest operas, at the age of 79. Good health, a happy home life, and the knack of avoiding heavy administrative burdens may be the keys to mathematical longevity. I leave it to historians and psychologists to determine the best years of a mathematician's life and why some mathematicians retain their creativity throughout their lives.

The long history of mathematics, fascinating as it is even when restricted to the unvarnished truth, has accumulated some anecdotal myths to add extra color. Here are a few, in chronological order.

The Egyptians used rope in the form of a 3–4–5 triangle to form the right angle at the base of a pyramid.

In a geometry book published in 1993, you will find this assertion: "To make a right triangle, the ancient Egyptians took a rope with twelve equally spaced knots in it, and then bent it in two places to form a triangle with sides of lengths 3, 4, and 5, as in this picture."

It is true that the Egyptians had rope. If you visit the Egyptian wing of the Metropolitan Museum in New York, you can see a sample for yourself. It looks just like modern rope. Moreover, wall paintings show that Egyptian surveyors used a rope the way modern surveyors use a long tape measure. It is also true that the 3–4–5 triangle has a right angle and that the pyramids have right angles.

However, there is no evidence that the Egyptians used this rope trick to make a right angle. The belief that they did goes back to a guess that the historian Moritz Cantor offered in his four-volume *Lectures on the History of Mathematics,* published in 1907:

> The Egyptians had carpenter squares, one being clearly shown in a wall painting of a carpenter's workshop. But the accuracy of this tool had to be assumed, and this seems to require some geometric construction. And presumably, in the case of such a solemn occasion as the founding of a temple, accuracy was checked anew. To be sure, the ritual for the temple founding does not disclose how this was done.
>
> Let us postulate without any foundation that the Egyptians were aware that when three edges of lengths 3, 4, and 5 are joined to form a triangle, the two shorter edges enclose a right angle. Now let us postulate that they divided a rope of length 12 by means of knots into pieces of lengths 3, 4, and 5. It is clear that this rope, when stretched around three stakes, creates a right angle.

Cantor never stated his claim as a fact. Indeed, his reasoning is full of several assumptions. But with the passage of time the assumptions vanished and the guess evolved into a historical fact.

Historians advise me to disregard any historical claim not accompanied by clear-cut evidence, namely, the original source. That suggestion appeals to me as a mathematician, since I like to see the assumptions and the logical steps laid out cleanly and clearly. That is why I don't read historical fiction. I enjoy history and fiction but not their mix, for then I have no idea what world I am in. I would as soon dine on a bowl of bean soup and ice cream.

The golden ratio appears in the Great Pyramid of Khufu, the Parthenon, paintings, the UN building, the dimensions of the most beautiful rectangles, the human body, and Virgil's Aeneid.

The *golden ratio* is the number $(1 + \sqrt{5})/2$, whose decimal begins 1.618. Dozens of articles purport to show that it is the key to the design of many beautiful objects. However, George Markowsky in "Misconceptions about the Golden Ratio," methodically disposes of each of these myths. I will give just a brief glimpse of what he found.

Though the ancient Greek mathematicians had met the golden ratio, there is no evidence that they connected it to aesthetics. Rather, it described a certain way of dividing a line segment into two pieces. There is a unique point C on the segment AB such that the ratio between AB and AC is the same as the ratio between AC and CB, as in the following picture. For that point C, the ratio AB/AC equals the ratio AC/CB and is called the *golden ratio*, a name that first appeared in 1835.

$A \qquad\qquad\qquad\qquad C \qquad\qquad B$

Does the golden ratio show up in the Great Pyramid? There are so many dimensions—width, height, slant height, edge length, half of width, and so on—that you could squeeze just about any number you wish out of them, just as you could find almost any letter in a bowl of alphabet soup. The proposal that the golden ratio shaped the design of the Great Pyramid was first made in 1859 and was based on a bogus translation of Herodotus.

Similarly, the claims that the golden ratio shaped the Parthenon, the paintings of Leonardo da Vinci, the UN building, the Aeneid, and the human body are without any support. In each case, there are so many numbers available that you could make just as strong a case for 2 or ¾ or any number you prefer.

Perhaps most of us have often heard that the most beautiful rectangles are those whose length is 1.618 times their width, that is, where the ratio of length to width is the golden ratio. There is no evidence for this claim. When Markowsky asked persons to choose the most beautiful rectangle from an array of rectangles of various shapes, the one chosen most often had a ratio of length to width of 1.83. You may conduct the experiment yourself, if you still believe in the golden ratio.

Archimedes cried "Eureka," and claimed he could move the earth.

We often hear that when Archimedes discovered the principle of buoyancy he leaped out of his bath and ran naked through Syracuse crying *Eureka,* which is Greek for "I found it." Maybe he did, and maybe he didn't. There is no mention of this in his surviving writings, some of which have chatty introductions. The episode is mentioned by Vitruvius, a Roman architect who lived in the first century B.C., some two centuries after Archimedes, who died in 212 B.C.:

> When he went down to the bathing pool he observed that the amount of water which flowed outside the pool was equal to the amount of his body that was immersed. Since this fact indicated the method of deciding whether the king's crown was made of gold, he did not linger, but, moved with delight, he leapt out of the pool, and going home naked, cried aloud that he had found exactly what he was seeking. For as he ran he shouted in Greek, eureka, eureka.

To me, it sounds like an embellishment invented long after the fact. I feel that Archimedes made so many exciting discoveries that he wouldn't run around town every time he had a fresh insight. But maybe the myth is true. In any case, I take it with a grain of salt (appropriately, a Roman proverb).

The first record of Archimedes' claim about moving the earth is in Plutarch's *Lives,* written three centuries after Archimedes: "Archimedes declared 'If there were another world, and I could go to it, I could move this one."

This assertion seems to acknowledge hearsay as its source. Maybe there were earlier written sources that have disappeared. However, I am surprised that the first surviving reference appeared so long after Archimedes' death. His accomplishments are so astonishing that his life needs no fictional additions. You need only read his treatment of floating bodies, in which he founds the science of naval architecture, to be convinced. Just to give a hint of what he discovered, consider the parabolic-shaped solid pieces of wood in the following figure.

Imagine that each of them is resting on a table. If you tilt the one on the left, it will return to its resting position. But if you tilt the one on the right, it will fall over. Archimedes discovered which parabolic

shapes are shallow enough so they return to their resting positions and which are too tall and will fall over.

Newton invented calculus to solve the problem of planetary motion.

Isaac Newton (1642–1727) invented calculus in 1665 and 1666 and did not tackle the problem of determining the planetary orbits until around 1680. When he did, he used a geometric approach, not the algorithms of his calculus. Some historians claim that he first analyzed the orbits with the aid of calculus, but because few of his readers knew the subject, he rephrased his arguments geometrically. There is no evidence for this in his surviving manuscripts, even though hundreds of pages survive.

Galois first wrote up some of his great discoveries the night before the duel in which he lost his life.

This myth goes back to a passage in E. T. Bell's *Men of Mathematics,* published in 1937:

> All night he had spent the fleeting hours feverishly dashing off his scientific last will and testament, writing against time to glean a few of the great things in his teeming mind before the death which he foresaw would overtake him. Time after time he broke off to scribble in the margin "I have not time; I have not time," and passed on to the next frantically scrawled outline. What he wrote in those desperate hours before the dawn will keep generations of mathematicians busy for hundreds of years.

It is true that Galois was killed in a duel and that the night before it he wrote several letters, annotated some of his papers, and on one of them wrote, "I have not time," but he wrote this only once. The rest of Bell's description is pure fiction, as Tony Rothman in his article "Genius and Biographers: The Fictionalization of Evariste Galois" demonstrates: "Galois had been submitting papers on the subject [of group theory] since he was seventeen. . . . During the course of the night he annotated and made corrections on some of his papers." What indeed would keep mathematicians busy had been written down from one to four years before this last night.

Gauss measured the angles between three mountains to see whether space is Euclidean.

I had heard for years that C. F. Gauss (1777–1855) had measured the sum of the angles formed by three mountain peaks to see whether their sum is really 180°. If the sum were different, then space would not obey the customary rules of geometry and therefore would be an example of a non-Euclidean space. For instance, M. Kline, in his *Mathematical Thought from Ancient to Modern Times,* wrote, "He found that the sum exceeded 180 degrees by 14.85 seconds. The experiment proved nothing because the [possible] experimental error was much larger." Kline even cites a particular page in Gauss's 1827 paper where the data appear.

It is true that Gauss carried out such an experiment, but he did so only to check whether the fact that the earth is not a perfect ball (being flatter at its poles) would seriously affect the calculations in his geodetic survey of Hanover. Arthur Miller, in "The Myth of Gauss' Experiment on the Euclidean Nature of Physical Space," notes that Gauss never claimed that his experiment was related to the nature of space. Miller conjectures that the myth originated in response to Einstein's general theory of relativity, published in 1916, when "The question of whether physical space is curved or not took on new meaning." Since Einstein used mathematics developed by Georg Riemann (1826–1866) in 1854, which was related to Gauss's 1827 paper, someone may have "extrapolated back (without reading, of course)."

As a boy, Einstein was poor in arithmetic.

People tend to think of Einstein (1879–1955) as a mathematician. He was not. He did not invent any mathematics; rather, he applied existing mathematics to the physical universe. It is more accurate to call him a "theoretical physicist" or a "mathematical physicist."

Was he weak in arithmetic? Not at all. On the contrary, he was quite good at it. The myth that he was poor in it is due to a quirk. When he was a pupil, his school inverted the grading system, so that what previously was a high grade became a low one. Anyone reading Einstein's report cards, but not knowing about this shift, would conclude that Einstein had suddenly lost his mathematical ability.

There is no Nobel Prize in mathematics because a mathematician who might have won the prize had an affair with Nobel's wife.

It is often said that the mathematician Gösta Mittag-Leffler (1846–1927) had an affair with Alfred Nobel's wife, and that he would have been a prime candidate for the prize. This myth appeals to the pride of mathematicians, but there is one catch. Nobel never married. Consequently, he never had a wife. To put it in other words, he was a lifelong bachelor. So much for the myth. It would be fascinating to find the name of the practical joker who invented the story. There is no evidence that Nobel disliked Mittag-Leffler.

If by questioning any of these myths I have deprived any readers of their favorites, I apologize. I believe the naked truth of history is already exciting enough. To read a carefully documented history book, one that includes the letters and diaries of the times, is a joy. Reading it, I feel as close as I can be to the events and the principals and that no one is coming between me and the past. I suppose that is why people collect old newspapers, reenact a Lincoln-Douglas debate or the battle of Gettysburg, touch the very plane that dropped the atomic bomb on Hiroshima, or cherish a chunk of concrete from the Berlin Wall. They want to bypass the "layers of historians" that come between them and the past.

Come to think of it, I like mathematics for the same reason I like to take my history straight. When a theorem lies on a piece of paper on my desk, nothing comes between it and me. There is no chance for a spurious myth to form and take on a life of its own. What I see is what is there. That is a rare experience these days, when everything seems to come with a spin.

··· 7 ···

The Rapid
Idiots

The automobile of today is basically the automobile of a century ago: four wheels, a motor, an accelerator, a brake, a steering wheel, and headlights. There have been pleasant embellishments, but an antique car travels side by side with the latest model on the same roads and highways. I suspect that the automobile of a century from now will resemble the automobile of today. But this is not the case with computers. No one can predict with confidence, even a few years ahead, what magic computers will be performing for us. Practically the moment you buy a new computer, it becomes obsolete.

A computer has two fundamental features. It is remarkably fast, with some computers able to carry out over a billion operations per second. Also, it is almost error-free. The combination enables it to free our minds from routine mental tasks, just as the steam shovel and the vacuum cleaner free our bodies from routine physical tasks. In a mere half century, computers have become so common that we are hardly aware of their presence. Perhaps the key date is even more recent—1977—when the Apple personal computer made its debut.

When I stop to think about it, I see that computers have slipped into my life in more ways than I had suspected.

- When I start my car, a computer adjusts the mix of air and gas.
- When I phone to ask for the balance in my bank account, a computer speaks to me.

- At the grocery store or gas station I don't need to use real money, just a plastic card, again thanks to some computer.

- When I need cash, I slide a plastic card into a slot in a wall and out come $20 bills even when my bank is closed, even on a holiday, even when I am thousands of miles away from home.

- As I watch the logo of a television channel appear on the screen, twisting, turning, expanding, I am seeing animation implemented by a computer (with the aid of lots of algebra).

- When I go bowling, not only does the automatic pinsetter pick up and replace the fallen pins—that's old stuff—but a computer, hooked to the pinsetter, does all the scoring, marking the spares and strikes, and enters the running total.

- When I make a long-distance telephone call, invisible computers quickly route it away from saturated circuits, record the time, number called, the place, and the charge. Later they print out the bill and, for all I know, drop it in the mailbox.

- Sitting in a chain restaurant, I assume that its temperature is controlled by a thermostat in the building. Yet it turns out that a computer at the national headquarters some two thousand miles away is in charge.

- When trying to find a report on a paper by the mathematician Roger Apéry, written around 1960, I first checked several years of the *Mathematical Reviews* in vain. Then the reference librarian told me that the *Reviews* is available on a CD-ROM (Compact Disk, Read Only Memory). In a few seconds, it displayed a list of all of Apéry's papers covering a period of four decades.

- Every so often I have used a computer in my research to carry out computations that would take years or decades if attempted by hand. The output sometimes refuted a conjecture or suggested a new one. In this respect, the computer serves me the way a telescope serves an astronomer, or as a microscope serves a biologist, enabling me to see further into the universe of mathematical phenomena.

When I was studying drunk versus sober driving, I had made a list of the number of accidents and the amount of traffic for each of the 168 hours of the week. I then wanted to divide the number of accidents by the amount of traffic to get a measure of accident risk for each of the 168 hours. Even with a calculator, this would have been a tedious task. However, a spreadsheet did the 168 divisions in a split second.

I can even use a computer to type papers full of mathematical symbols and complicated expressions. With the aid of special software, I can typeset an equation with the same ease as typing a sentence, symbol after symbol on a line. For instance, when I type $\$\${3 + 2\backslash over\{1 + 7\}\} = 0.625\$\$$, the printer will display

$$\frac{3+2}{1+7} = 0.625.$$

Since no compositor comes between me and the final book, there is little chance for new errata to creep in. I am in a position to produce camera-ready text and pictures.

Having hooked my computer by a modem to the outside world, I can write a mathematician in Hungary and receive a reply the same day. Airmail, alias "snail mail," has a turnaround time of at least three weeks.

To write this book I use a computer as a word processor. Actually, I write the first draft with pen and ink, dipping the pen frequently into the ink bottle, as if I were using an eighteenth-century quill. (An engineer once advised me to "use the device with the fewest moving parts.") Then I transfer the text to the computer, where the editing is miraculous. If I switch two paragraphs, I don't have to retype a page, and if I delete a paragraph, I don't have to retype a whole chapter, as was the case in the bygone era of the typewriter. However, I must remain on guard and resist the temptation to consider that there is nothing left to do. The display on the screen looks neat and final, no matter how much work is left. Moreover, as Joshua Stein warned his fellow lawyers in *The Practical Lawyer,* "A computer can certainly help you to be more productive, but it can also lead you to focus on details (and endless fussing with format) rather than big-picture issues."

But hard disks fail, a hand strikes the wrong key, power goes out, computers crash for no explicable reason—and the work of days can be lost. A computer is in what physicists call "unstable equilibrium." It is like an egg balancing on an end: The slightest deviation from perfection may trigger disaster. (I am thinking of a colleague who lost all the grades in a class accumulated over a semester—without a floppy disk copy or so-called hard copy of a printout on paper.) For this reason, the personal computer user must back up regularly, at least every computing day, sometimes three times a day. Large organizations may have a second computer duplicating the work and memory of their main computer. A "mission critical," such as the moon landing, may have three identical computers doing the same computations.

As we add more and more computers and computer-operated devices in our home, our car, our office, or factory, the problem of maintenance grows. We are given an 800-number to call for service, but it may be down for the weekend or offer us only the wisdom of a busy signal. There is an upper limit to the number of devices—mechanical or electronic—that we can keep in working order. When I think of the many devices in my home—the dishwasher, garbage disposal, thermostat, washing machine, dryer, microwave, television, radio, clocks, furnace, air conditioner, and so on—it is a wonder that I don't spend all my time responding to a breakdown. Indeed, there have been days upon days when I was calling repair people to rush over. Sometimes I am tempted to get rid of all the machines, including my computer, and live a simpler life, one not so vulnerable to minor shocks.

Though computer scientists speak of the artificial intelligence of a computer and have written software that defeats a good chess player, we should keep in mind that a computer is just a bunch of switches that open and close incredibly quickly. It is no smarter than an eggbeater. Sometimes, when I use the spell-check, I begin to think the computer does have a mind. However, when upon not finding the word *Leibniz* in its dictionary, it offers *Albion* as the correction, I am reminded once again of its profound stupidity.

A computer does what it is told to do. What comes out of it is determined by what goes in. As programmers warn, "garbage in, garbage out." The forecasters who say, "The computer predicts," are really saying, "I made some assumptions, defined a few variables, gave them some numerical values, and cranked the computer." The assumptions and the program determine what the computer will spit out. In short, "opinion in, opinion out." As one economist warned, if you can't do the arithmetic on the back of an envelope, a computer will not be of any help.

One example will illustrate why we should be on guard against forecasts made with the aid of a computer. It concerns the overbuilding of hotels in the 1980s resulting in a glut in the 1990s. In an article titled "Causes of Hotel Industry Distress," M. J. and J. J. Flannery wrote,

> Although technology has advanced at a rapid pace . . . individuals remain fallible. The use of computerized models to forecast future cash flows led many to the misconception that investment analysis is an exact science.

Computers made it possible to produce numerous scenarios. This ability to play "what if" games should have improved judgments. Instead it led to the acceptance of pseudo-scientific certainties.

Computer models rarely focused on downside assumptions. Many analysts "massaged the numbers," adjusting inflation, strengthening occupancies, pushing rates slightly, and shaving expenses. When projects did not yield the desired results, their advocates reworked the numbers until the models provided the desired returns.

I pay no attention to reports of "computers predicting" unless I see all the assumptions on which they are based. In other words, I want to speak to the ventriloquist, not the dummy whose jaw moves up and down in imitation of thought.

Computers are also infiltrating classrooms at all levels, permitting pupils to conduct numerical and geometric experiments too cumbersome to do with paper and pencil. But there is a danger here, especially in public schools, which are supported by taxes—the unmentionable T-word. Without an adequate support system on the spot, a breakdown in the computers could undermine the best-laid lesson plans. Too many computers languish for lack of personnel trained to exploit or service them.

There is also the risk that a computer may play the role of the autocratic teacher who says, "Don't ask why. Just do it the way I say." It is then a mysterious black box, and using it places the pupils in a position of dependency. This circumstance is most unfortunate in mathematics, where pupils should develop self-confidence and accept nothing on faith. Several teachers have told me that "the capacity to reason seems to get lost when you start pressing keys too early."

The ability of computers to store and access massive amounts of data moves power to organizations that can accumulate information about individuals. This can be good: A policeman, stopping a speeding car, can immediately check the driver's record. Or bad: A group may gather so much information about a person—shopping pattern, credit history, travel, and so on—that it amounts to an invasion of privacy.

As I think of all the wonderful labor- and mindsaving inventions of the industrial and electronic revolutions, two questions perplex me. Why do so many of us still work so hard and yet have so little spare time? And are we happier than our parents and grandparents? We are more comfortable, and it is easier to take care of the daily chores, but which of the "improvements" are in fact obstacles to a good life? What are the hidden costs we pay unwittingly, costs that we may someday discover, when the final reckoning is done, exceed the more conspicuous, well-advertised benefits?

I think of these questions when I read the diary my mother kept in 1903 in Dubuque, Iowa, when she was 13. Here is a typical entry: "Went to school. Took violin lesson. In evening went to piano lesson. Practiced. Embroidered cushion for living room sofa." It suggests a way of life when people of necessity were more self-reliant and had more time to explore and develop their inner resources.

Could we be climbing on an Escher staircase, thinking that we are going up with each step, only to find to our surprise that at the end we are back where we started or, perhaps, even lower? Being an optimist, I like to think that the answer is no. Still, the question is worth pondering, not just for computers.

··· 8 ···

The Mother of Invention

According to an old Latin saying, "Necessity is the mother of invention," but it is just as true that "Curiosity is the mother of invention." It was curiosity, not necessity, that compelled Michael Faraday to explore electricity and magnetism in the early part of the nineteenth century. When asked, "What is the use of all this?" he replied, "What is the use of a newborn babe?" He was not thinking about the telegraph, the telephone, electric lights, radio, television, radar, compact disks—the very apparatus of our world. He was driven by the urge to answer the fundamental questions common to physics, chemistry, biology, and mathematics: "What is the nature of the universe in which we dwell?" or, more generally, "What is true?" The hard-won answers to these questions compose the treasury of our civilization, increasing our options and magnifying our power to act, whether to our benefit or to our harm.

It is difficult, perhaps impossible, to predict the long-term impact of our actions, especially of new discoveries. When IBM announced the invention of the transistor in 1947, people saw it as a device for use in hearing aids, nothing more; they certainly did not see it as the key to making computers smaller. In 1949, IBM estimated the world demand for computers would be satisfied by 15 machines. When Charles Townes and Arthur Schawlow invented the laser in 1958, they were not thinking of scanners in grocery stores and libraries, compact disks, precision measurement, eye surgery, or fiber optic cables.

Many discoveries that lie entirely within the world of mathematics also turn out later to have surprising applications in the "real world." They are inspired by curiosity, by the compulsion to answer a question, to explore the unknown.

I will sketch three examples. To fill in the details, you may consult the references at the end of the book.

••• Knots •••

In the 1880s physicists believed that light traveled on a substance called ether, which they assumed filled every nook and cranny of space. It was conjectured that an atom was simply a tangled knot made of ether, with different elements corresponding to different knots.

Let's pause a moment to describe a mathematician's view of a knot.

To make a knot, take a piece of string, tangle it up, then glue its two ends together. Following are four pictures of knots.

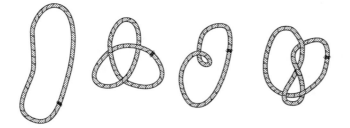

The knot on the left, which isn't at all tangled, is called the *unknot*. The second knot cannot be untangled no matter how you may move the string about. It really is different from the unknot. At first glance, the third knot may look tangled, too, but a little pulling will untangle it to reveal that it is the unknot in disguise. The fourth picture shows another knot that cannot be untangled. Moreover, it can be shown to be really different from the second knot.

Inspired by the physics, mathematicians began to study knots. Even after Einstein showed that there is no need for an ether, they continued their research. Here is the basic question they asked: "Given two diagrams of knots, how can we tell whether they really depict the same knot?" After a century's labor, there is still no automatic procedure for deciding, though some complicated algebraic tools settle the question in certain cases.

If anything would appear to be of no practical use, surely it would seem to be the theory of knots. Yet in the 1980s this theory turned out to be of aid in unraveling the chemistry of DNA molecules. These molecules, which we usually think of as having the shape of a spiral staircase, or helix, often shrivel up into the shape of a tangled knot. The theory of knots helps analyze the properties of these forms of DNA. Knot theory also has recently been applied in statistical mechanics, a branch of physics.

••• Probes •••

Back in 1917, the Austrian mathematician Johann Radon (1887–1956) raised a seemingly useless question, which I will phrase in terms of a fruit-cake.

A fruitcake contains cherries, candied pineapple, nuts, raisins, and orange peels scattered through the cake. It is the unfortunate marriage of what might have been a charming fruit salad with a tasty cake. A bit of brandy or rum is tossed in to preserve the concoction, which tends to appear around Christmas in the form of a gift. Since few of its recipients are fond of fruitcake, it is sometimes saved to be passed on as a gift the following year. Consequently, some fruitcakes may be several hundred years old, as perfectly preserved in their alcohol as an ice-age bison in the Siberian tundra.

In any case, imagine that you want to know exactly where each of the ingredients is without cutting the cake open. You may pass an extremely narrow hollow tube, as thin as a spider's thread, through the cake. You then weigh the material collected in the probe. Imagine that you can do this in every possible direction and every part of the fruitcake. The question Radon asked is, "If I know exactly how much each probe weighs, will I have enough information to figure out the location of each fruit and to tell what it is?"

An orange peel is made of a very light matter. So is a walnut. A piece of pineapple is much denser. Even though it may be of the same size as a walnut, it weighs more. It has a higher density. Assume that each of the different fruits in the cake has a different density, so each fruit can be identified by its telltale density.

Radon showed that, theoretically at least, using only the weight of each needlelike probe, he could figure out indirectly what was inside the cake and its location. As the title of his paper indicates, "On the Determination of Functions from Their Integrals along Certain Manifolds," he was not thinking of anything as practical as analyzing a fruitcake. Browsing through this paper you would see many x's and y's and a host of symbols used in calculus. Anyway, there was no way to make such probes without ruining the cake. His entire investigation was simply an attempt to satisfy his curiosity.

Yet decades later, Radon's discovery was to be applied in astronomy, molecular biology, geophysics, optics, and medicine. For example, computerized tomography (CT) enables a doctor to look inside a patient without recourse to surgery. An X-ray beam plays the role of the thin hollow tube. The amount of the beam absorbed as it passes through the patient plays the role of the weight collected in the tube. The data from thousands of such beams are processed by a computer in order to construct a cross-sectional map. This process requires so many calculations that only an electronic computer could carry them out quickly enough for them to be of any use. Since at the time Radon wrote his paper there were no electronic computers, he certainly did not anticipate any application of what was to become famous as the *Radon transformation*. Once again, curiosity was the mother of invention.

••• Codes •••

My third example comes from the theory of numbers, traditionally viewed as the branch of mathematics least likely ever to be applied. As G. H. Hardy put it in 1940, "No one has yet discovered any warlike purpose to be served by the theory of numbers . . . and it seems very unlikely that anyone will do so for many years."

I suspect that he would have been disappointed to learn that in 1977 three mathematicians, R. Rivest, A. Shamir, and L. Adleman, using number theory, invented a new type of secret code, which they published in a paper called "A Method for Obtaining Digital Signatures and Public-Key Cryptosystems." This code was not their first attempt. Rivest and Shamir had proposed 42 codes, and Adleman had cracked them all. But he could not crack the 43d. This code provides a secure and secret way for banks and businesses to transmit information. It is sufficiently warlike that the federal government considered publishing the code and the technique for breaking it to be a violation of the 1954 Munitions Control Act.

I will describe only the number theory on which the code is based and direct you to the References for the details.

First, recall a few definitions from arithmetic. The *whole numbers* are 1, 2, 3, 4, . . . , the numbers used in counting. One whole number, D, is said to *divide* another one, N, if there is a whole number Q (for "quotient") such that D times Q is N. For instance, 3 divides 12 since $3 \times 4 = 12$. If D divides N, then we also say that D is a "*factor* of N." As an example, the factors of 12 are 1, 2, 3, 4, 6, and 12.

When D is a factor of N, we also say that "N is a *multiple* of D." For instance, the multiples of 3 are 3, 6, 9, 12, 15, . . . , and so on, counting off by 3s. When N is a multiple of D, we also say that "D goes into N evenly"—with no remainder.

A number that has exactly two factors, itself and 1, is called a *prime*. The first few primes are 2, 3, 5, 7, 11, 13, 17, 19, and 23. (Note that 1 is not a prime since it has only one factor, itself.)

The next concept we need is "raising to a power." If N is a whole number, we shall denote $N \times N$ as N^2, read as "N to the second power" or as "N squared." Similarly, $N \times N \times N$ will be denoted as N^3, read as "N to the third power" or as "N cubed." More generally, if e is a whole number, we will denote the product of "e of the Ns" by N^e, read as "N raised to the eth power." For example,

$$5^2 = 5 \times 5 = 25, \quad 5^3 = 5 \times 5 \times 5 = 125, \quad \text{and} \quad 5^4 = 5 \times 5 \times 5 \times 5 = 625.$$

In the eighteenth century, Euler discovered a remarkable property of these powers, of which the following is a special case. Take any two distinct prime numbers, p and q. For any whole number N, form the number

$$N^{(p-1)(q-1)+1} - N.$$

Euler proved that this number is always a multiple of the product of the two primes, $p \times q$.

To bring this down to earth, take the simplest case, when p is 2 and q is 3. Euler then asserts that for any whole number N,

$$N^{(2-1)(3-1)+1} - N.$$

is a multiple of 2×3. This boils down to the assertion that

$$N^3 - N$$

is a multiple of 6, for every choice of the whole number N. Let's check it when N is, say, 4. In that case, $N^3 - N$ becomes $4^3 - 4$, which is $64 - 4$, or 60. As Euler predicted, 60 is indeed a multiple of 6. You may easily check it for other choices of N, such as 1, 2, 3, and 5. You may also want to check Euler's assertion for another pair of primes, such as 3 and 5. If you use larger primes, the arithmetic becomes onerous, even with a calculator.

Euler would be astonished to learn that his assertion that a certain number is a multiple of the product of two given primes would become, two centuries later, the basis of a secret code.

To use the code, a bank, say, chooses two large primes p and q. By large, I mean on the order of 75 digits long. It then computes their product $p \times q$, which is about 150 digits long, and makes this product, called the *key*, available to its customers. However, it does not reveal what the two primes are. (An article by Martin Hellman in the August 1979 issue of *Scientific American* shows how the key is used to make a code.)

To crack the code, a sleuth would have to find the two primes whose product is that gigantic key. That means finding the factors of a 150-digit number. Now, finding the factors of 12 can be done in a few seconds by hand, but finding the factors of a much larger number could take the fastest computers a very long time. To get a feel for the difficulty, find the two primes whose product is 1,739, which is a fairly small number. Then imagine the trouble of finding the factors of a 150-digit number; it is not an appealing prospect.

In 1977, it was thought that a 150-digit number would be large enough to hide its two prime factors for centuries. In that year, *Scientific American* published a short message based on a 129-digit key and offered a prize of $100 to anyone who could decipher it, presumably by finding the two prime factors of the published number.

Years passed without anyone deciphering the message. In 1990, RSA Data Security, the company founded by the three inventors of the code, published several smaller challenge keys, the three smallest having 100, 110, and 120 digits. Mathematicians managed to find their factors, but the one with 129 digits resisted all attacks until 1994. Six hundred volunteers in 24 countries, each working at one or more personal computers, cooperated over a period of eight months to uncover the two prime factors. (If these 600 volunteers had done the calculations by hand at the rate of one per second, the decoding would have taken over 5 million years.) By the way, the deciphered message turned out to be hardly earth-shaking; it was "The magic words are squeamish ossifrage." The group donated the prize to the Free Software Foundation, which distributes software programs free of charge.

For the present, codes based on 150-digit numbers still seem to be secure. But with computers getting ever faster and with new techniques for faster factoring being invented, even such a number may not be large enough.

The three cases—knots, the Radon transformation, and Euler's theorem—show that mathematics done to satisfy curiosity can later be put to practical use. The list of examples could go on for pages, but I mention just two more, for good measure. Complex numbers, which became an accepted part of mathematics early in the nineteenth century, were the perfect tools for analyzing alternating current at the end of the century. A type of geometry introduced in the nineteenth century turned out to be exactly what Einstein needed in the early part of the twentieth century to express his general theory of relativity.

But there is no need to extend the list, since the three examples make my point: "Curiosity is the mother of invention." Mathematicians may do mathematics because they find the questions intriguing and the discoveries surprising, profound, eternal, and beautiful. Society supports them because those discoveries often have great practical value in ways that no one, not even the discoverer, could foresee.

··· 9 ···

What Is a Job, Really?

W hat is the first question that comes to mind when a headline reads, "Army Base to Close," "Injunction against Logging," "Merger," "Gambling Casino Proposed," "New Prisons Needed," "Higher Tax on Tobacco," or "Clergy Pleads to Reduce Christmas Shopping"? Answer: How many jobs will this create or destroy? Usually the first or second paragraph of the report offers the numbers.

There is a good reason for this underlying concern about jobs. Those who don't have a job worry about finding one; those who do have a job worry about losing it. It's hard to concentrate on the truth and beauty of mathematics if you are wondering how to put bread on the table. For this reason, the next chapter examines the role mathematics plays (or doesn't play) in the many ways of earning a living.

First, though, let's pause briefly to analyze the notion of a job.

As surely as water is made of hydrogen and oxygen, a job consists of two separate and distinct parts: a *production station* and an *income station*. Everyone needs an income station in order to be able to buy the necessities and luxuries of life. For an individual, this is the key part of a job. Society as a whole needs a certain number of production stations in order to supply those goods and services.

There is no reason whatsoever why the number of income stations needed will happily equal the number of production stations needed. (Contrast this with the balance between the number of males and the number of females, maintained by whether an X or a Y chromosome happens to pass to the next generation.) Indeed, after two centuries of the industrial revolution and a century of the electronic revolution, with their countless laborsaving and mindsaving inventions, it would come as a surprise if there still were enough production stations to go around. The persistent unemployment of at least 11 percent in Western Europe and 6 percent in the United States suggests that an advanced nation will experience an imbalance. The International Labor Organization estimates that worldwide "almost one in three workers—or 820 million out of 3 billion— is either without a job or under employed."

As long as we think of a job as a single undivided unit, we will be as handicapped as a chemist who isn't aware that water is made up of two elements. When we worry about jobs, we are really worrying about a lack of income stations.

There may be little relation between the size of the income station (the money earned) and the value of the production station (what is produced). For instance, a person who teaches the young or cares for the old occupies a small income station but a crucial production station. On the other hand, an executive in a cigarette company may occupy a gigantic income station, but the production station may, according to the Surgeon General, kill 1,000 Americans a day. You have probably noticed many such perplexing contrasts.

An income station may even have no associated production station— for example, a welfare recipient or an inheritor of a family fortune. Similarly, a production station may have no associated income station—for example, a hospital volunteer or a museum docent. The Brittons of West Jordan, Utah, are a specific illustration; they find good homes for hundreds of abandoned dogs free of charge. *Newsweek* and CBS recognize the almost invisible people who occupy valuable production stations that have only empty or small income stations. As the president of *Newsweek* wrote, "Why are only movie stars and other celebrities deemed worthy of the spotlight? What about an award show for ordinary Americans who have helped people in their community or performed unsung acts of heroism?" That makes sense.

The gradual shortening of the workweek has tended to maintain an approximate balance between the two types of stations. But if a nation does not realize there is a fundamental imbalance, it may not try to eliminate it. Instead, it may be left to new inventions, new fads and diversions, or more intensive advertising—all of which agitate an economy—to create enough production stations to meet the demand. This haphazard approach, which depends on the working of an "invisible hand," may not do the trick. No wonder that concern about jobs is so prevalent.

The next chapter describes in detail the production stations in the United States and the number of workers in each type. That will help us keep our feet on the ground when we wonder, "Who needs what mathematics?"

··· 10 ···

What's in It for Me?

It is possible to lead a happy and useful life at home or in the workplace without ever adding two fractions or playing with an algebraic equation. I am not going to claim that everyone must study calculus in order to get a good job. I will not try to persuade anyone to become a mathematician. Instead, I will simply describe the level of mathematics needed in specific occupations, together with the number of people working in that occupation. I will not try to predict which occupations will expand and which will shrink, nor which will need more mathematics and which less. The future is hard to predict, and I leave that task to *Occupational Outlook Quarterly,* published by the Bureau of Labor Statistics, and its survey, *The American Work Force: 1992–2005.*

Disregard jobs for the moment and consider the question, "What mathematical skills does a person need to cope with daily life?" If you have a friend, a spouse, or an accountant to do all the arithmetic for you, then you don't need to do any of it yourself. But if you don't want to yell for help every time you cut a recipe in half or figure out what 18 percent interest on a credit card means, or if you want to be able to recognize that your calculator has displayed a ridiculous answer, then skill in dealing with fractions and percentages is a must. Moreover, as I will show, this level of mathematical expertise is all that is required in most jobs.

However, when we see computers creeping into every corner of our lives, we may well wonder whether the need for even arithmetic will disappear. A laser scans a bar code, a computer enters the amount the customer pays and takes care of the rest, even the sales tax and the change, with the coins falling automatically down a chute.

But what happens if the computer crashes? Or if the human being who programmed the computer entered the wrong price? I have seen cashiers helpless in such a calamity.

Though most jobs use only the arithmetic that gets a person through daily life, I am not going to suggest that people should stop studying mathematics at the sixth grade. I believe in opening doors to the full range of opportunities and keeping them open as long as possible, not in closing them. Many of the better-paying and less repetitive occupations do require mathematics beyond arithmetic.

Sometimes the mathematics requirement sneaks in indirectly. For example, Sacramento City College, which offers many two-year vocational programs, requires that students pass an algebra examination to graduate. One of the admission prerequisites at the University of California is three years of high school mathematics, and students are "strongly advised to take college preparatory mathematics each year in high school." Or a law school may advise potential applicants to study "mathematics and logic," among other areas of science and the humanities.

Sometimes, although the mathematics requirement is spelled out, it still may come as a surprise. At Sacramento City College, a student of electronics must take second-year algebra and trigonometry; a future nurse, a year of algebra. Some students majoring in psychology at the University of California must take a year of calculus.

According to a poll by Louis Harris and Associates, "Many high school students are making decisions about whether to continue their studies of mathematics with little or no guidance from parents, teachers, or school counselors. Over a third of the students who planned to stop taking mathematics as soon as possible were interested in studying science in college. This paradox was especially strong among minorities."

Moreover, studying a subject is usually not enough to master it. There are two dependable ways to achieve mastery: One is to teach it, even as a tutor; the other is to take the course that follows it, which applies the subject. Algebra helps review arithmetic, trigonometry illuminates geometry, and analytic geometry reinforces both algebra and trigonometry. Calculus not only reviews all of these areas and ties them together but also offers a wide variety of applications.

I was reminded of the value of a follow-up course when an engineer I know was hiring an assistant to help with some calculations. He had put in a help-wanted ad for someone who had "two years of college, including a year of calculus."

"Are they really going to be using calculus?" I asked.

"No, but that way I can be sure that they know trigonometry. They're going to be doing a lot of trig calculations."

He pointed out that most engineers do not use calculus in their work. They needed calculus when they were in college to understand their physics and chemistry courses, which built the basis for their engineering studies.

An article in the *New York Times*, "Stepping into the Heels of Their Career Moms," mentioned the case of a girl, age 13, who was debating the importance of algebra with her mother, a managing director with J. P. Morgan. "In my career I'm never going to use algebra." Her mother flashed a V for victory sign, glad that her daughter "had totally accepted that having a career was a given." I hope that the V did not also signify "no need for algebra." If the girl wants to follow in her mother's footsteps, she may choose to major in business administration or economics in college. If so, she's in for a shock: Both usually require a year of calculus—and skill in doing algebra quickly and accurately is one of the keys to mastering and enjoying calculus. Or what if she decides to become a doctor? Once again, a year of calculus.

I am not implying that everyone will be studying calculus. As we will see, only 4 million out of 121 million workers require calculus at some point in their careers, even though they may not apply it in their job.

What is the best way to find out what mathematics is required in which occupations? That is the question I faced as I thought about writing this chapter. I certainly was not going to poll thousands of randomly chosen workers. Luckily, I was not the first person to consider the question.

Hal Saunders's book *When Are We Ever Going to Use This?* lists the mathematics used in 100 occupations. He asked people "which of some 60 math topics they used, and how they used them." For instance, he found that 32 occupations use the Pythagorean theorem. The poster that summarizes the data Saunders accumulated should be displayed in every classroom where mathematics is taught. It makes a quick and convincing case for the importance of mathematics.

However, I wanted to paint the whole picture, covering all occupations. To do this, I consulted *The Complete Guide for Occupational Exploration*, which lists the requirements of over 12,000 occupations. To find how many people were employed in each occupation, I had to look at a different source, *The American Work Force: 1992–2005*, which divides the workforce into some 500 categories. Then I had to reconcile the data from the two sources, since they often used different groupings and titles.

The *Dictionary of Occupational Titles,* which lists over 20,000 types of jobs, helped merge the two books. As I read it, I was amazed by the variety and complexity of the careers needed to sustain a modern economy. It was fun to browse through the index and try to guess the meaning of such job titles as "twister-doffer," "bull-gang supervisor," "roustabout," "long-chain beamer," "gimp tacker," "germ drier," "priller," and "bottle-house pumper."

The *Complete Guide for Occupational Exploration* is organized around 12 major interest areas, such as *artistic* and *scientific.* It breaks these into 66 work groups and then into 348 subgroups; these subgroups are broken into over 12,000 occupations. For each occupation, it lists the level of reading, language, and mathematical development required. I looked only at the mathematics requirement, which comes in six levels, from 1 to 6, with 6 being the highest. Here they are:

Level 1 Add and subtract two-digit numbers. Multiply and divide 10 and 100 by 2, 3, 4, and 5. Perform the four operations of arithmetic with coins as parts of a dollar. Operate with cup, pint, quart; inch, foot, yard; ounce and pound.

Level 2 Work with fractions and decimals. Compute ratio, rate, percentage. Draw and interpret bar graphs.

Level 3 Compute interest, discount, profit, loss, commission, markup, ratio, proportion, percentage. Find areas and volumes. In *algebra,* work with variables and formulas, polynomials, square roots, radicals. In *geometry,* work with plane and solid figures, properties of angles, and pairs of angles.

Level 4 In *algebra,* deal with basic functions (linear, quadratic), solutions of equations, and inequalities. Work with logarithmic, trigonometric, and inverse functions, limits and continuity, probability, and statistical inference. In *geometry,* the axiomatic approach, plane and solid, and coordinates. In *shop math,* practical applications of fractions, percentages, ratio and proportion, measurement, algebra, geometric construction, trigonometry.

Level 5 In *algebra,* work with exponents and logarithms, mathematical induction, binomial theorem, permutations. In *calculus,* differentiation and integration of algebraic functions. In *statistics,* frequency distributions, normal curve, analysis of variance, correlation, chi-square, sampling theory, factor analysis.

Level 6   In *advanced calculus,* limits, continuity, implicit function theorem, differential equations, infinite series, complex variables. In *modern algebra,* groups, rings, fields, linear algebra. In *statistics,* experimental design, statistical inference, econometrics.

For convenience, I'll translate these levels into what I think are the roughly corresponding courses:

Level 1:   rudiments of arithmetic (grade 4)

Level 2:   fractions and decimals (grade 6)

Level 3:   arithmetic of business, some algebra (first-year algebra)

Level 4:   algebra, trigonometry, geometry (second-year algebra, semester of trigonometry, a year of geometry)

Level 5:   some analysis, calculus, statistics (a year of calculus, a semester of statistics)

Level 6:   the core of a college-level mathematics program (three years of calculus, a semester of abstract algebra, a semester of statistics)

As I remarked earlier, level 2 represents the arithmetic useful in daily life. Only with level 3 does algebra enter the scene. This level is important for occupations where the workers may deal with formulas expressed in symbols or may solve simple equations. Level 4 approximates the entrance requirements of the University of California. A high school curriculum usually can take a student to level 4 or 5. Level 5 is roughly the mathematics in a biology, premed, or economics major, while level 6 corresponds to majors in physics, engineering, or mathematics.

Trigonometry appears in level 4 and calculus in level 5. I'd like to say a word about the terms *trigonometry* and *calculus.* These words frighten some people, even those who go on to major in mathematics. *Trigonometry* is just Greek for "three-angle-measure," and *calculus* is Latin for "pebble," once used in calculating. Perhaps renaming them "three-angle-measure" and "pebble" might demystify them. Any students who have honed their algebraic skills by adequate practice so that they are quick and accurate will find these later courses perfectly understandable—the natural next steps. Some 100,000 high school students take the advanced placement test in calculus each year, and some 750,000 college students are taking a calculus course. So "pebble" could not be so mystifying.

Reading through *The Complete Guide*, I noted the mathematics levels assigned to the careers. These were fairly constant within each of the 348 subgroups and even in some of the 66 groups. I summarized these in some 70 categories that came close to the titles in *The American Work Force*. The following table is the result, providing an overview of the role of mathematics in the workplace. For more specific information about a particular occupation not found in the table, I refer you to the two sources I used and to the *Dictionary of Occupational Titles*. The Further Reading section at the back of this book lists additional guides to the world of work, some covering just one city, some describing salaries, others commenting on the work environment. I was astonished by the amount of advice, guidance, and information available. An information freeway already runs through our lives. It is made of paper.

By the way, don't take each figure literally but rather as a ballpark estimate. The populations I give come from the Bureau of Labor Statistics and are based on information supplied by employers. The Census Bureau, using a survey of 60,000 households, also estimates the same populations. The two figures never agree, though they are often quite close. Incidentally, the total workforce in 1992 was about 121 million.

Occupation	Number of People	Level of Mathematics
Executive, administrative, managerial	15 million	
Managers, administrators, executives (in finance, construction, food service, lodging, education, property, etc.)	11 million	4
Management support (accountants, auditors, cost estimators, tax collectors, loan officers, etc.)	4 million	4
Professions	17 million	
Engineers	1.7 million	6
Architects, surveyors, life scientists	400,000	5
Computer, mathematical, operation research systems (actuaries, system analysts, statisticians, etc.)	800,000	6
Physical scientists (chemists, physicists, meteorologists, geologists, etc.)	400,000	6

(continued)

Occupation	Number of People	Level of Mathematics
Social scientists (economists, psychologists, urban planners, etc.)	400,000	5
Social, recreational, religious workers (clergy, human services, recreation, etc.)	1.1 million	4
Lawyers and judicial workers	800,000	4
Teachers (exclusive of math and science)		
Preschool through grade 6	3.3 million	2–4
Special Education	62,000	2–4
Secondary	620,000	4
Postsecondary	780,000	4
Other (adult, vocational, etc.)	1.3 million	3
Librarians (curators, restorers)	230,000	3
Counselors	240,000	3
Health diagnosing (physicians, dentists, etc.)	880,000	5
Health assessment and treatment (registered nurses, therapists, pharmacists, etc.)	2.5 million	3–4
Writers, artists, entertainers (dancers, athletes, designers, etc.)	2.1 million	1–2
Technicians and related support	4.4 million	
Health and engineering technicians and technologists (licensed practical nurses, electronic technicians, etc.)	2.8 million	3–4
Other technicians (computer programmers, aircraft pilots, legal assistants, etc.)	1.6 million	3–4
Marketing and sales	14 million	
Cashiers	2.9 million	2–3
Salespersons (retail)	3.9 million	2–3
Marketing and sales supervisors	3.7 million	2
Other sales (insurance, real estate, etc.)	3.5 million	3
Administrative support, including clerical	20 million	
Information clerks (receptionist, hotel desk, ticket agents, etc.)	1.6 million	2

Occupation	Number of People	Level of Mathematics
Adjusters, investigators, collectors	1.2 million	2
Stock clerks	1.8 million	2
Traffic, shipping, receiving clerks	820,000	2
Bookkeeping, accounting, auditing clerks	2.1 million	2
Secretaries, stenographers, typists	4.2 million	2
General office clerks	2.7 million	2
Bank tellers	530,000	3
Clerical supervisors and managers	1.3 million	2
Other administrative support and clerical	4 million	2
Service	16 million	
Janitors, cleaners, pest control	3.3 million	1–2
Waiters and waitresses	1.8 million	1–2
Food counter, fountain, and related	1.6 million	1
Other food preparation and service	1.3 million	2
Nursing and psychiatric aides	1.4 million	2
Other health service	650,000	2
Personal service (child care, cosmetologists, home health aides, etc.)	2.3 million	2
Private household workers	870,000	2
Fire fighting	310,000	2
Law enforcement	980,000	2–3
Guards	880,000	1–2
Other protective service	230,000	2
All other service	500,000	2
Agricultural, forestry, fishing, and related	3.1 million	
Gardeners and groundskeepers	880,000	2
Farm operators and managers	1.2 million	2
Farmworkers	850,000	1–2
Other	170,000	2
Precision, production, craft, and repair	14 million	
Construction trades (carpenters, electricians, plumbers, etc.)	5.3 million	3

(continued)

Occupation	Number of People	Level of Mathematics
Supervisors of mechanics and repairers	270,000	3–4
Vehicle and mobile equipment mechanics and repairers (cars, buses, trucks, aircraft, etc.)	1.9 million	3–4
Electrical and electronic equipment repair	630,000	3–4
Other mechanics, installers, repairers	1.4 million	3
Printing, textile, wood precision workers	500,000	2
Precision production (metalworkers, assemblers, etc.)	500,000	2
Operators, fabricators, laborers	19 million	
Numerical control, machine tool operators, metal fabricating and processing, plastic molding, etc.	2 million	4
Printing, binding, textile, woodwork machine operators, etc.	3.3 million	1
Hand workers (electronic assemblers, welders, cannery, etc.)	2.5 million	1
Truck drivers	2.9 million	1–2
Bus drivers	650,000	2
Other transportation and material moving	1.6 million	2
Helpers, laborers, other material movers, by hand	5.2 million	1

A glance at the table suggests a few observations. Out of a total workforce of 121 million people, about 4 million are in occupations where the mathematics level is 5 or 6, which includes calculus. That is only 3 percent of the workforce. On the other hand, almost 80 million people earn a living with only level 1 or 2, which is arithmetic. That is about two out of every three workers.

As the economy grows more sophisticated, as computers and robots take greater responsibilities upon themselves, will the workforce need more mathematics or less? There are arguments both ways. I leave it to the *Occupational Outlook Quarterly,* published by the Bureau of Labor Statistics, to predict employment trends. In any case, students are tending to take more mathematics than in years past. According to the

Condition of Education, published in 1994 by the National Center for Education Statistics and based on data from the Department of Education, 56 percent of high school graduates in 1992 took two years of algebra. Ten years earlier, the figure was only 37 percent. In 1992, 70 percent took geometry; ten years earlier, 48 percent had. That means that in 1992 over half of high school graduates had reached about level 4 in mathematics.

As a rule of thumb, the income of an occupation is related to the amount of education, and that, in turn, is usually related to the level of mathematics. But income is only one factor in choosing a career. Some people may want to work outdoors. Some may want a 9-to-5 job that "can be left at work" so time and energy are available for other interests. In any case, it is clear that the more mathematics a person knows, the more choices that person will have.

Students have often said to me, "I like mathematics. What can I do with a degree in it, just teach?" Based on what I have noticed at my campus, about half the students who graduate with a bachelor's degree in mathematics do go into teaching. But the other half, all at level 6, have a wide variety of choices. Some become actuaries, system analysts, marketing specialists, network administrators, financial analysts, and so on. One became a Hollywood joke writer. Others find that their undergraduate training is a springboard to the professions, such as medicine and law.

A physician, Arthur Staddon, who had majored in mathematics, wrote, "Mathematics opened the doors to the very best medical schools. The discipline of analytical thought processes prepared me extremely well for medical school. In medicine one is faced with a problem which must be thoroughly analyzed before a solution can be found. The process is similar to doing mathematics."

Another mathematics major, Jonathan Blattmachr, who went on to become a lawyer, had a similar view. "Although I had no background in the law—not even one political science course—I did well at one of the best law schools. I attribute much of my success there to having learned, through the study of mathematics, and, in particular, theorems, how to analyze complicated principles. Lawyers who have studied mathematics can master the legal principles in a way that most others cannot."

Blattmachr was not the first to observe this phenomenon. Thomas Jefferson, in a letter to a fellow student in about the year 1765, wrote, "Mathematics [is] so useful in the most familiar occurrences of life, and so peculiarly engaging and delightful, as would induce everyone to wish an acquaintance with [it]. Besides this, the faculties of the mind, like members of the body, are strengthened and improved by exercise. Mathematical reasonings and deductions are therefore a fine prepration for . . . the law."

Of the 1.2 million bachelor's degrees earned yearly, only about 12,000, or 1 percent, are in mathematics. Clearly, those 12,000 students have many choices, even without continuing their studies in graduate school. Students who appreciate the beauty and precision of mathematics may specialize in it in college and then later put it to the more worldly purpose of earning a living. In short, they can have the best of both worlds—the aesthetic and the practical. By studying mathematics, they demonstrate that they can carry out a sustained chain of reasoning. Combining their major with a little computer science, statistics, physics, and biology puts them in a flexible position when looking for a job.

Perhaps the best way to summarize is to quote from the *Occupational Outlook Quarterly*:

> Deciding how much high school math to take is easier if career goals have been established. However, it is better to take what may seem to be too much math rather than too little. Career plans change, and one of the biggest roadblocks in undertaking new educational or training goals is poor preparation in mathematics. Furthermore, not only do people qualify for more jobs with more math, they are also better able to perform their jobs.

··· 11 ···

The Action Syndrome

Chapter 12 offers a quick overview of the history of attempts to reform the teaching of mathematics in the last hundred years. But to understand these reforms, perhaps any reforms, we must first pause to discuss the *action syndrome*.

The action syndrome helps a person cope with the stress of action. It reduces several options ultimately to one. It enables the doer of an act, the "actor," to commit to this one option, to suppress doubts, and to sustain dedication. Whether the action is wise or foolish, the action syndrome permits the actor to focus on pursuit of the goal.

The syndrome is an automatic response to a challenge, helping the individual to act in spite of fear and uncertainty. It changes a person of half a mind or of two minds to a single-minded actor.

The action syndrome, a type of self-hypnosis, is necessary to complete a sustained act. Originally it may have served only as a survival mechanism, when decision concerned whether to pursue, to hide, or to flee, where hesitation meant death. But even in the modern world it is critical as it aids the transition from indecision to decision, from the hesitant "I think I shall do it" to the firm "I shall do it." Commitment and the action syndrome are inseparable.

The syndrome polarizes the mind as a magnetic field aligns iron filings. This polarization is described by such words as stubborn or persistent, close-minded or dedicated, monomaniacal or tenacious—the choice depending on whether you oppose or endorse the action.

Up to the moment of commitment, a person is free to weigh risk and gain, to imagine all the pitfalls, and to consider alternatives. But the thinker, in order to become the doer, must fix upon a single course or nothing can be accomplished. The action syndrome permits this switch to occur—from observer to actor—as real as the change of a caterpillar into a butterfly.

A conversation between Xerxes, a Persian king of the fifth century B.C., and his advisor, Artabanus, as reported by Herodotus, reveals the role of the action syndrome:

> Artabanus said, "It is best for men, when they take counsel, to be timorous, and imagine all possible calamities, but when the time for action comes, then to deal boldly."
>
> Xerxes replied, "Fear not all things alike, nor count up every risk. For if in each matter that comes before us you look to all possible chances, you will never achieve anything. Success attends those who act boldly, not those who weigh everything. Great empires can only be conquered by great risks."

A bodybuilder can experience the power of the action syndrome. As John Lee Brown, Jr., put it, "There are two parts to the human mind, the conscious and the subconscious. The conscious mind is able to make judgments and assessments. If it sees something is impossible, it will tell you. But once you've told your subconscious mind something, it takes you at your word and makes things happen. I have told my mind I can do anything I want to, and I can."

The moment of transition from noncommitment to commitment marks a critical change, yet it is little understood. Freud touched on it when he wrote, "When making a decision of minor importance, I have always found it advantageous to consider all the pros and cons. In vital matters, however, the decision should come from the unconscious, from somewhere within ourselves. In the important decisions of our personal life, we should be governed by the deep inner needs of our nature."

Facing the gap more directly, President Kennedy admitted, "The essence of ultimate decision remains impenetrable to the observer—often, indeed, to the decider himself. . . . There will always be the dark and tangled stretches in the decision-making processes—mysterious even to those who may be the most intimately involved."

Once the act is begun, the action syndrome sustains the actor. Each new obstacle strengthens rather than weakens the resolve, so the actor, whether an explorer, an inventor, or a reformer of mathematics education "carried away" with a sense of mission, persists. The president of Continental Airlines, when ending low-fare service, remarked on this phenomenon, "It started as a pilot project that should have been proven before it was expanded. But once this thing started rolling, it was awfully hard to turn it around."

Nothing is so persuasive as the force of conviction. Those who cannot convince themselves certainly cannot convince anyone else. The action syndrome therefore enables the actor to draw others to the cause.

As we recount the history of reforms in the teaching of mathematics, keep the action syndrome in mind. It will help make sense of the past and perhaps even guide us in the future.

··· 12 ···

Where Have All the Reforms Gone?

A battle has raged throughout this century over how mathematics should be taught. One side has emphasized computational skill, the other, understanding. The pendulum has swung back and forth—from "back to basics" to the "New Math" and "problem solving"—without settling on a peaceful compromise. Often this battle flows over into the "Letters to the Editor" section of newspapers when the school authorities are about to choose a new mathematics curriculum. Only an attempt to ban a book from the school library or disband the football team can arouse such an intense passion.

Perhaps if mathematics were taught in the ideal way, with just one pupil in a class, the conflict might never have started. I could see the advantage of "one teacher, one pupil" instruction when I took up oil painting and asked Kati, an art major, to teach me the basics. Right off she had me stretch a canvas, squeeze paints onto a palette, and pick up a brush. "Well, what do you want to paint?" she asked. I had brought along a photograph of a path through woods down to a lake. "Fine," she assured me. "Go ahead." It was as simple as that. As I tentatively put paint on canvas we discussed choices and goals. And so I learned.

That's one way mathematics should be taught, tailored to each pupil, who serves an "apprenticeship." Then the teacher would know where the pupils stand, their interests, the appropriate motivation. But economics rules out this approach. Instead, a secondary school mathematics teacher typically deals with 150 pupils, which amounts to about a quarter hour a week per pupil, hardly time to start a conversation. An elementary school teacher faces a class of some 30 pupils. Breaking it into small groups, where the pupils can talk to each other, can at least help preserve a human scale.

That a teacher deals with groups rather than individuals affects the whole curriculum, not just mathematics. However, in the eyes of the public, *mathematics* is practically synonymous with *school.* I suspect that if you asked most people what subject comes to mind in free association with school, it would be mathematics. Mathematics is, after all, the most conspicuous subject, visible year after year from kindergarten through 12th grade. Moreover, it is a cumulative structure; for example, percentages depend on fractions, and fractions depend on the arithmetic of the whole numbers. That means that a pupil who does not understand a key idea may suffer the consequences years after. (This could happen even in the so-called spiral curriculum, which keeps cycling back to earlier concepts.)

No wonder, then, that in the debate on a constitutional amendment imposing one minute of silence in school, an editorial cartoon shows the interior of a church with the minister saying, "In addition to prayer in school, the government has mandated we set aside time to do math problems." Not spelling, not grammar, not history, but mathematics. A letter to the editor advised, "Anyone who thinks there's no school prayer has never taken a math test." Are tests in all other subjects so easy?

I recall a professor of education who had a penchant for illustrating bad teaching by citing examples chosen exclusively from mathematics. One day there happened to be a mathematics major in her class who asked, "Isn't there bad teaching anywhere else?" The question put a stop to her habit.

Perhaps bad teaching is more visible in mathematics than in other subjects because any confusion in learning it is easy to spot: Answers are usually either right or wrong. For this reason, we should expect to find an eternal discontent in the way mathematics is taught. If you browse through *The Mathematics Teacher,* the main journal devoted to instruction in mathematics, you will find constant lamentation, going back to its first volume in 1908, where one teacher wrote, "One of the most obvious facts about mathematics in our schools is a general dissatisfaction." The tone in 1911 was even less cheery, "Our conference is charged with gloom. I have attended funerals, but I do not remember a more mournful occasion than this. We are failures and our students are not getting anything worthwhile."

Year after year, the complaints in *The Mathematics Teacher* persist. I will skip ahead to 1958, when we read, "The traditional curriculum is meaningless, and by heading for abstract mathematics the modernists are moving further from reality." This was an early warning about the group developing what came to be called "the New Math." More about that reform later. Still, in 1994, the University of Chicago School Mathematics Project complained, "The student today still encounters a variant of the elementary school curriculum designed for the pupil of a hundred years ago."

In the pages of *The Mathematics Teacher* one finds many plausible explanations for the discontent: too much routine computation, too much theory, not enough applications, too many applications, not enough use of calculators and computers, too much use. And, among others, not enough attention to mathematics as an art, as part of culture. Often someone blames poorly prepared teachers, as in this typical complaint, dated 1910, "All the modern physical equipment cannot replace the able teacher, nor can the most excellent text book take the place of the living word."

As a last resort, blame the parents or the whole society, as is done in this excerpt from 1911, pointing to "the distractions of modern life. One of the sources of the distraction is the weakness of too many American parents, and little cooperation with the school." Nothing has changed by 1992, when we hear, "Before the United States can begin meaningful educational reform, we Americans must decide what we expect of our students. Do we place a high value on academic achievement? Or do we give other goals, such as being popular or a good athlete, a higher priority?"

In response to the criticisms, the twentieth century has witnessed numerous reforms—some as small as a model classroom, others at the school or district level, and a few as large as a city, state, or the nation. These reforms spring forth even though there is no agreement on the cause of the problem. It is as though a doctor keeps plying patients with a variety of pills without ever figuring out what ails them.

As the pendulum swings from one magic solution to another, each decade gets its name: The "back to basics" 1950s; the "New Math" 1960s; the "back to basics" 1970s; the "problem-solving" 1980s; and the "group learning" 1990s. What makes a splash depends a good deal on the passion of some articulate, dedicated speaker willing and able to spread the gospel of reform at conferences or on the availability of grants from foundations or the government. Someone has to be convinced that he or she has found the key to successful mathematics education. Then the action syndrome takes over: Nothing is more convincing than a convinced prophet. Looking back at old reforms, I get the impression that the prophets are constantly reinventing the flat tire.

This admonition about reforms, written in 1909, holds today: "To start a reform requires sufficient energy to overcome the inertia of custom. The history of the teaching of mathematics shows a series of fluctuations from one emphasis to another, and all reform carries a good deal which in time is set aside as worthless. Some reformers see little or no good in anything of the present, but think all should be upset and an entirely new situation built upon new foundations. True reform seldom, if ever, came in this way."

Keeping these words in mind, I will describe four reforms, one small, one medium, the big one that produced the New Math, and, finally, the jumbo one of today, dubbed by its proponents "the *Standards*" and by its opponents, "the new New Math."

In 1929, the superintendent of schools in Ithaca, New York, asked his fellow superintendents to find what could be cut from the curriculum in order to make room for the newly mandated subjects "such as safety, health, and thrift instruction." Responding to this challenge, L. P. Benezet, the superintendent in Manchester, New Hampshire, wrote back, "It is nonsense to take eight years to get children thru ordinary arithmetic. The whole subject could be postponed until the seventh grade, and could be mastered in two years by any normal student."

Benezet had already been criticized for dropping practically all of arithmetic from the first two grades. However, he felt that if his letter "represented my real belief, I would be falling down on the job if I failed to put it into practice." Over a period of several years he then conducted experiments that he reported in three extensive articles in the *Journal of the National Education Association* in 1935 and 1936.

He knew that he could count on the cooperation of the children and teachers. But what about the parents? How many parents would let their children serve as guinea pigs in such a risky experiment? Though he sent out notices to the parents telling what he planned, he got no protests. Luckily for him, "not one parent in ten in the districts spoke English as their mother tongue. Had I gone into schools where the parents were high school or college graduates, I would have had a storm of protest—and no experiment."

Here is what happened. In Benezet's words, "The 6th graders were divided into two groups. The experimental group had no arithmetic until beginning the sixth grade and the traditional group had it starting in the third grade. At the beginning the traditional group excelled. By April the two groups were on a par. In less than a year the experimental group had been able to attain the level of accomplishment which the traditionally taught children had reached after three and a half years." Moreover, the pupils were shown the reasons behind the processes, such as, "why a correct answer is obtained in the division of fractions by inverting the divisor and multiplying."

What did he do with the time saved? He put it into "reading, reason, and recite." The "experimental" children developed more interest in reading, a better vocabulary, and greater fluency of expression than the pupils who came from homes where English was spoken.

Whenever I read his articles, I feel that I am right there with him, that I am participating in his experiment. Though the action syndrome begins to work upon me, I know that it will be impossible to spread his message. Too many parents speak English. And so it is that his reform disappeared from the stage, leaving scarcely a trace. In home schooling or small private schools where pupils proceed at their own pace, not an imposed one, one can see Benezet's theory confirmed.

The second reform I recount has left more of an impact on the way mathematics is taught. I know it well because I was one of its two ringleaders.

When my son Joshua was in high school in 1968 I became familiar with the text he was supposed to read. The exposition and exercises were horrible—pedantic, abstract, and tedious. I started to visit the class conducted by Cal Crabill, a highly respected and experienced teacher, to see how I might help out. After a while, Cal and I decided that the students should move their seats around and work in groups of four. That way they could help each other, get immediate feedback, and practice speaking the language of mathematics. The teacher still had a critical role, but it was a different one from that of only lecturing. Instead, he would go about the room, checking the progress of each group. In case of a general confusion, the teacher would address the entire class. Also, the teacher would introduce a topic and summarize it after the class had worked on it. In short, pupils would learn sometimes in small groups, sometimes in lectures.

The enthusiastic response of the students encouraged us to expand the experiment through the whole algebra-geometry-trigonometry sequence. As with Benezet, the action syndrome took hold of us, and we spread the gospel through talks and panels at mathematics teachers' conventions.

Throwing out the text, we mimeographed suitable material, characterized by an informal style and by exercises that invite experiments and discussion. Eventually these notes became three textbooks, with the geometer Don Chakerian joining us in the geometry text. These were the first mathematics texts built around a "small-group learning" approach.

Now, some 20 years later, *cooperative learning* has become the fad of the 1990s. But when I visit a classroom I often see the technique misused: Either the teacher, feeling that all should be left to the pupils, says nothing, or else, at the opposite extreme, answers questions too soon, before the students in the groups have had time to think. That is not what Cal and I intended. Our books remain in print, little advertised, lost beneath the welter of newer books sporting four-color illustrations. In any case, our innovation survives, but hardly in a form we recognize. This shows me that there is no "teacher-proof" text. The teacher will always remain far more important than any text or computer program.

The third reform I describe is the School Mathematics Study Group, known in its time as SMSG, and dubbed the "New Math." It was a product of a quirk in the Cold War.

On October 4, 1957, the Soviet Union launched *Sputnik 1,* the first man-made satellite, which was viewed by millions of Americans as it passed overhead. I watched it, a small bright dot, very high, moving faster than an airplane. Immediately the cry went up, "The Russian rockets are better than ours. We are perilously behind. Our educational system is at fault. We must reform our mathematics and science instruction at all grades."

In response, the National Science Foundation funded SMSG, backing it for several years with millions of dollars as it developed materials to serve as models for commercial texts.

What is still not so well known is that the United States had the means to put up a satellite before *Sputnik.* However, President Eisenhower was reluctant to employ a military rocket in a peaceful enterprise; he thought it would send the wrong message. As he wrote later, "The separation of the [peaceful] earth satellite program from the military missile program had disadvantages, the principal one being that the satellite program could not make full use of all the advances made in . . . military missiles. The Army . . . could undoubtedly have placed a satellite in orbit some time late in 1956, considerably before the Soviets." We were able to launch our first satellite, *Explorer 1,* on January 1, 1958, just three months after *Sputnik.* Had Ike been less scrupulous, there might never have been an SMSG.

In view of the importance of SMSG, let's look back and see what it did and how its contributions were received.

At the start of the project in 1958, its head, Ed Begle, announced five principles, which seem sensible enough:

1. No one can predict exactly which mathematical skills will be useful in the future.

2. No one can predict exactly what career a student will choose.

3. Teaching which emphasizes understanding without neglecting the basic skills is the best for all students, whatever their ability, and makes the best preparation for any vocation that uses mathematics.

4. An understanding of the role of mathematics in our society is essential for intelligent citizenship.

5. Any normal individual can appreciate some of the beauty and power of mathematics and this appreciation is an important part of a civilized person's cultural background.

He also described a reasonable way to proceed: "SMSG will combine the efforts of mathematicians, teachers, and teachers of teachers. The resulting materials should fuse correct mathematics with sound pedagogy."

Most of the work of SMSG was accomplished during the four years from 1958 through 1961. Everything was well thought out: Professors and teachers worked together writing the texts, which were tested in hundreds of classes. On the basis of the feedback from teachers and students, the teams then revised the exposition. The authors certainly had their feet on the ground.

No one can say that teachers were not well represented on the writing teams. In the first writing session in the summer of 1958, there were 16 teachers out of the 47 participants; in 1959, 41 out of 106; in 1960, 49 out of 101; and in 1961, they were a clear majority, 40 out of 71.

Right from the start, SMSG gathered strong support, as is shown by these comments in *The Mathematics Teacher* of 1959:

> Curriculum reforms have been advocated before. But this movement is different. College people and secondary-school teachers are sitting down together to draw up materials.

And, in 1961,

> Those of us who are familiar with what appeared to be promising efforts over the last 50 years realize that success is often less than is anticipated. Perhaps the closer involvement of teachers at local levels may be the difference which will ensure the success of this program [SMSG].

In spite of its sensible assumptions, the care it spent on developing new materials, its willingness to revise its texts, and the enthusiastic endorsements, there were early signs of unease.

In 1960, in the same journal, a teacher, Wallace Manheimer, expressed serious misgivings:

> Fellow teachers, do you feel pushed around? Do you feel that your pupils are cheated by having such a reactionary teacher as yourself? That they are learning mathematics that is 600 years old? Surely the newspapers have been telling you so.
>
> "Modern mathematics" is all the rage. The movement to teach it has become the latest panacea for the ills of our subject. Unlike other movements of the past it is fortified by considerable university backing, widespread publicity, and large sums of money.
>
> The classroom teacher has the right to disagree with the fashionable prediction that the secondary mathematics curriculum will be altogether different in twenty years. May he guess that the trial of the new ideas will force a retreat upon its proponents.

But storm clouds were also blowing in from the colleges. A memorandum signed by 65 mathematicians appeared in the journal in 1962:

> It would be a tragedy if the curriculum reform should be misdirected and the golden opportunity wasted. There are, unfortunately, forces which may lead us astray.
>
> Mathematicians, reacting to the dominance of education by professional educators who may have stressed pedagogy at the expense of content, may now stress content at the expense of pedagogy and be equally ineffective. Mathematicians may unconsciously assume that all young people should like what present day mathematicians like.

Undeterred by such warnings, SMSG continued on its way. In 1963, it had to publish *A Very Short Course in Mathematics for Parents,* which acknowledged

> Parents are discovering that something has happened to the mathematics courses their children are studying. Homework assignments contain new words and ideas, and parents often find that they can no longer help their children do their arithmetic problems.

One of those new ideas that dumbfounded parents was *bases other than ten.* Because of its symbolic importance in the New Math, I'll take a moment to introduce it. This notion, which was at least three hundred years old, is neither harder nor easier than our customary way of writing numbers using only the symbols 0, 1, 2, 3, 4, 5, 6, 7, 8, 9. What makes it seem difficult is that we are not used to it. It is another language, a different way of writing numbers. What is confusing is that we use the usual symbols of our decimal (base ten) arithmetic to describe it. I'll sketch just enough of the details to give you some idea of how it works.

We have ten fingers. Numbers less than ten we denote by special symbols, 0 through 9. After that, we need no more symbols. Instead, we group by tens, and then by tens of tens (called "hundreds"), and so on. When we write 21, we mean "two tens and one," which we shorten to "twenty-one." When we write 201, we mean "two hundreds, no tens, and one."

There is nothing sacred about the number ten. If we had only four fingers, that is, two on each hand, we might count by fours instead of tens. (There would be no reason to count by tens.) In this case, we would need only the symbols 0, 1, 2, and 3 to describe the numbers less than four. This is the world of base four, in contrast to our world, where we are used to base ten.

In the world of base four, the dots in the following figure would be viewed as "two groups of four and three left over."

This number would be written 23. Of course, we are so used to our ten fingers that we might read it as "twenty-three" and think that there are many more dots, namely, "two groups of ten plus three." But base four is a different language, and though its words may look like ours, they have different meanings. In that language, "23" would be read as "two fours plus three," and "10" simply as "four." It takes time to "forget we ever saw base ten" and become used to writing numbers in other bases and seeing what their arithmetic looks like—probably a few days.

But just to give a hint of what a child living in the base four world would see, consider its multiplication table. Because there are only the four symbols 0, 1, 2, and 3, that table is a delight, having only nine products to memorize.

×	1	2	3
1	1	2	3
2	2	10	12
3	3	12	21

The table certainly looks strange, but only because we all grew up speaking the base ten language. For instance, to find 3×3, the child would draw this figure:

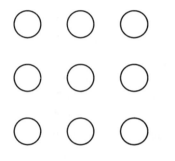

and then count it off by fours, as in the following figure.

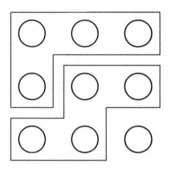

Since the figure is made up of two groups of four and one left over, the child would write it as 21 and so find that $3 \times 3 = 21$. (In order not to mix this up with base ten, read it as "two fours and one," not as "twenty-one.")

If we had only one finger on each hand, we might use base two and need only the symbols 0 and 1 to record the smaller numbers. The multiplication table the child would need to memorize is simply the one at the right.

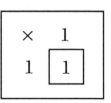

If our children ever heard about this, they might revolt against base ten and demand that we use base two. However, they might later regret their choice when they learn that base two has a drawback: It requires many digits to write even small numbers. For instance, in base two, 2 is written as 10, 4 is written as 100, 8 as 1000, and our 16 as 10000.

On the other hand, base two makes a computing machine happy. Since there are only two digits, 0 and 1, it can easily represent them electrically as "current off" or "current on."

I don't want to turn this chapter into a lesson on other bases. To do a complete job—showing how to add and multiply—would take a few pages. The reader would have to spend a few hours practicing to get the feel of the arithmetic. I just wanted to show the challenge that faced pupils, parents, and teachers when they dealt with other bases than the one they had grown up with. Maybe teachers should learn to work in another base just to be able to empathize with their pupils as they learn to work with base ten. In any case, the best way to learn the usual base—ten—is to work with it. There is no need to get it all mixed up with other bases.

SMSG believed it had to educate teachers about different bases, as its advice to seventh-grade teachers illustrates, "This unit deepens the pupil's understanding of the decimal notation for whole numbers. Since in using a new base, the pupil must look at the reasons for carrying and the other mechanical procedures in a new light, he should gain deeper insight into the decimal system."

Richard Feynman (1918–1988), winner of the Nobel Prize in physics, had a less sanguine view of the use of bases other than ten, as he recounts in his book *Surely You're Joking, Mr. Feynman:*

> In the early sixties my assistant, Tom Harvey, said, "You oughta see what's happening to mathematics in the school books! My daughter comes home with a lot of crazy stuff."
>
> Many people thought we were behind the Russians after Sputnik and some mathematicians were asked to give advice on how to teach math. . . .
>
> I'll give an example: They would talk about different bases of numbers—five, six, and so on— . . . That would be interesting for a kid who could understand base ten—something to entertain his mind. But what they had turned it into, in these books, was that *every* child had to learn another base! And then the usual horror would come: "Translate these numbers which are written in base seven, to base five." Translating from one base to another is an *utterly useless* thing. If you *can* do it, maybe it's entertaining; if you *can't* do it, forget it. There's no *point* to it.

By 1965, SMSG had completed most of its work, and a book reviewing its accomplishments appeared: *SMSG: The Making of a Curriculum,* by W. Wooton. Its tone was optimistic: "The reception thus far from students, teachers, teachers of teachers, parents, administrators, and mathematicians has been quite gratifying and indicates its influence will last. Its first four years have shown that classroom teachers and ivory-towered mathematicians can work together."

Yet in a few years most of the innovations of the New Math disappeared. Perhaps it went too far in introducing technical terms, abstraction, and sophisticated exposition. Perhaps the teachers weren't trained to deal with it. Perhaps it never was given a chance. For instance, a sales representative for one of the commercial versions of SMSG advised teachers, "If you don't want to do the New Math, just omit the first hundred pages."

But not all was in vain. SMSG gave the number line a lasting prominence in the classroom. As anyone who has looked at a thermometer or ruler knows, numbers can be viewed as naming points on a straight line. Whole numbers, fractions, even negative numbers occupy points on this so-called number line, as shown in the following figure.

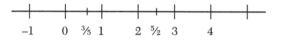

With the aid of the number line, pupils can see the world of numbers as a geometric object. In addition, SMSG brought probability into the curriculum and made it possible for many high school students to take calculus.

The sad fate of so massive and auspicious a reform as SMSG—one that combined the efforts of teachers and professors, collected feedback from the classroom, and revised its materials accordingly—did not put an end to major reforms, funded by private foundations or by state or federal governments. Instead of *Sputnik,* the stimulus might be the poor showing of American pupils on international examinations, or the belief that minorities need a different approach to mathematics, or the desire to incorporate calculators or computers.

Undaunted by lessons from the past, professors and teachers still dare to develop new texts. When I read the prospectus of one of these reforms I felt that it had everything right: its goals, its careful testing, the years of revisions, the feedback from pupils and teachers. I began to feel that I was part of their team, and the action syndrome seized hold of me, "At last someone has found the right approach. It is guaranteed to work." This is what the professors and teachers in SMSG must have felt when they were in the thrall of the action syndrome. But then I saw the final product—a high school geometry book over 800 pages long, requiring 12 pages devoted to the definition of a point, and I lost hope.

Another reform, even more extensive than SMSG, debuted in 1989. In that year the National Council of Teachers of Mathematics (NCTM) published *Curriculum and Evaluation Standards,* a book that describes in general terms what should be taught. Two years later, a companion volume appeared, *Standards for Teaching Mathematics,* advising how to teach and how to prepare teachers.

The intent of the *Standards* is "to establish a broad framework to guide reform in school mathematics in the next decade. It is a vision of what the curriculum should include, in terms of content priority and emphasis." Both books call for drastic reform, away from teacher telling to student discovery, away from routine computation toward conjecturing and solving nonroutine problems. Both books urge that instructors should persistently emphasize "doing" rather than "knowing," and that mathematical ideas should originate with the children rather than the teacher, in an inquiry-oriented manner. Like the New Math, it emphasizes understanding, but that is where the similarity ends.

The New Math grew out of the Cold War; the *Standards* is a response to the poor showing of American students on international examinations. The New Math focused on the logic of mathematics; the *Standards* focuses on the pupil, who is to "construct" mathematical knowledge out of experience. The New Math prepared sample textbooks; the *Standards* merely gives criteria for judging texts and teaching styles. The New Math was the work of mathematicians and teachers in roughly equal numbers; the *Standards* was developed almost exclusively by teachers. The New Math worked only on the curriculum; the *Standards* aims to overhaul curriculum, pedagogy, and the way students are graded.

Nevertheless, the goals of the *Standards* resemble those of the New Math: "Students should value mathematics, be confident with their ability to do mathematics, be able to solve problems, and communicate and reason mathematically."

To attain these goals, children will generally work in small groups to *construct* their knowledge. (The technical term is *constructivism.*) As the *Standards* quotes from *Everybody Counts,* "Effective teachers can stimulate students to learn mathematics. Students learn mathematics only when they construct their own understanding. They must examine, apply, prove, communicate. This happens most readily when students work in groups, discuss, present, and take charge of their own learning. Students do not learn simply what they have been shown."

The *Standards* faces the old conflict between computation and thinking: "Some proficiency with paper-and-pencil calculations is important, but such knowledge should grow out of problems that need them." Also,

"The calculator renders obsolete much of the traditional complex paper-and-pencil proficiency."

The *Standards* summarizes its reforms in a two-column table, from which I take a few entries.

Increased Attention	Decreased Attention
Open-ended problems	Routine one-step problems
Discovering mathematical ideas	Doing fill-in-the-blank worksheets
Reasoning from experience	Relying on teacher as authority
Connecting mathematics to the world outside	Developing skills out of context
Active involvement of students	Teacher exposition
Developing number sense	Memorizing rules

The *Standards* does not just set the tone for mathematics education. It spells out the criteria that determine which experimental curricula will be funded, which workshops will be held, and which texts will be published. Its objectives, like those of the New Math, are laudable.

Perhaps the new reform will succeed. As Jack Price, the president of NCTM, wrote in 1994,

> The success of the current reform movement will become evident as we proceed. There are a number of reasons why this is true. All segments of the mathematics education community—mathematicians, mathematics educators, teachers, and supervisors—are at last in the same book if not on the same page. Second, we have eliminated the "trickle down" syndrome that proved disastrous in earlier efforts; we have made certain that *everyone* is an equal partner in development and implementation. Third, the reform has a basis in research and in sound philosophy. Fourth, the publishers are creating materials that implement the *Standards*. Fifth, the technology is available to assist in the endeavor. Finally, governments support the direction of the reform.

Supporters of the new reform believe that it will succeed because it avoids what they see as the errors of the New Math. Elaine Rosenfeld, chair of the California Mathematics Subject Matter Committee, in justifying the selection of certain texts, said, "The New Math was totally abstract, used unfamiliar contexts, communicated poorly with parents, and did nothing to make math accessible to children. It furthered parents' belief that math was only for the gifted few. It failed to pay attention to how students learn. It did not help teachers understand the changes that were asked of them."

But just as the New Math met with some isolated skepticism, so has the *Standards*. Chester Finn, Jr., who helped prepare the Bush administration education proposal, wrote in *Education Week* in 1993, "Even in the faddy world of K-12 education, the *Standards* have met with rare acceptance. Seldom has so profound a change in conventional wisdom and standard practice had such homage paid to it, so little resistance shown to its onrush, so few doubts raised about its underpinnings. We better hope they got it right. If not, the lemming-like rush to follow its lead could find us hurtling off a precipice."

Zalcman Usiskin, a professor of mathematics education at the University of Chicago, was more specific in his misgivings. In a talk gently titled "What Changes Should Be Made for the Second Edition of the NCTM Standards," he offered some not-so-gentle criticisms.

> Dissent from the *Standards* has been meager, primarily because NCTM discourages any criticism of it. If one is not for the *Standards* one must be against good mathematics, good teaching, and good evaluation. . . .
>
> There is virtually no memory in the *Standards* of what has been recommended before and failed, and no indication of what if anything is truly new in the *Standards*. Many of its recommendations were never tested on a large scale. . . .
>
> Although support for change in mathematics education is based in great part on the low performance of U.S. students in international comparisons, the *Standards* have not taken the best ideas from what is done in other countries. The curricula elsewhere have been ignored. Why? One reason is that these curricula do not follow the philosophy in the *Standards*. Elsewhere they do not believe that children always have to construct knowledge for themselves."

The ideas in the *Standards* are not new. As far back as 1938 the Progressive Education Association suggested, "Students must deal with whole situations rather than practice upon specific skills. . . . Individuals or small groups may present conflicting conclusions on the same problem." In 1940, in the NCTM yearbook, we find, "There is a trend toward leading pupils into new topics through their own experiences."

In 1972, *Mathematics Framework,* which was supposed to guide mathematics education in California, advised, "The ideal classroom climate fosters the spirit of 'discovery.' It also provides ways for pupils to direct their own learning under the guidance of a curiosity-encouraging teacher. Self-directed learning requires pupil-involvement in creative learning experiments. That is not accomplished by 'Now, here is how it goes' lectures." But these recommendations had scarcely any impact. What makes us think the *Standards* will succeed in the whole nation if the *Framework* could not succeed in one state?

I am disturbed that the authors of the *Standards* do not cite any pilot project or any school district as a model to show that their goals can be achieved in the real world. That means that they are proposing to change the way an entire generation learns mathematics without checking the feasibility of their recommendations. A manufacturer introduces a new soap with more care, first testing its reception in a few stores or towns before committing to mass production.

No matter how desirable its goals and suggestions are, the *Standards* must be implemented by publishers and teachers. Out of curiosity I looked at some of the texts recently adopted in California, purporting to meet the criteria of the *Standards.*

Here is how one of the seventh-grade texts has the pupils "discover" that the sum of the three angles of a triangle is always 180°. Under the heading "Work Together," it has these directions:

> Begin with a paper triangle that is a different shape from those of the other members of your group. Number the angles of the triangle and tear them off the triangle. Place the three angles side-by-side so that pairs of angles are adjacent and no angles overlap.
>
> What seems to be true? Compare your results with those of your group. Make a conjecture.

On the very next page, facing these directions, we find, "In the Work Together activity you discovered that the following statement is true." In boldface follows, "The sum of the measures of the angles of a triangle is 180°."

What kind of discovery or constructing of knowledge out of experience is that? From what I know about pupils, they would glance at the bold type on the next page and stop experimenting.

If you really want discovery, you ask the pupils to draw several triangles, measure each of their three angles, and add them. (Incidentally, this gives some practice in addition.) Then the class could compare results and guess what might be true. They may, and they may not, conjecture that the sum should always be 180°, the size of a straight angle (twice a right angle). At that point the teacher might say, "Yes, it always is 180°." But that would deprive the pupils of a chance to learn mathematical reasoning.

Instead, the teacher could say, "Experiments suggest that the three angles add up to about 180°. But we're not sure whether their sum is always exactly 180. I will show you that the sum is really 180, without doing any experiments."

"First of all, there are a couple of things you need to know about angles. In this figure, which shows two parallel lines cutting a third line, angles x and y are equal.

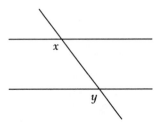

Also, in the next figure, angles y and u are equal.

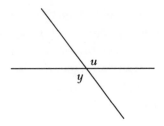

So, in the next figure, angles x and u are equal.

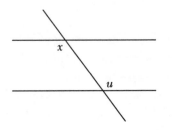

"Now imagine any triangle, with angles A, B, and C, as in the following figure.

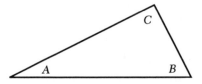

Draw a line through one corner parallel to the side not through that corner.

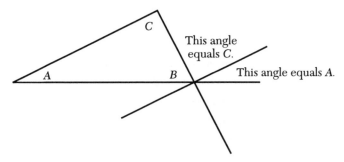

The three angles at the corner that lie above the horizontal line together form a straight angle, which has 180°. But, by what we had just noted about angles, these three angles are equal to the angles A, B, and C."

The teacher may then ask the class to experiment with the sum of the four angles in a four-sided figure. Even though the teacher gave them the full explanation for the triangle, there is much left for the pupils to explore and explain on their own.

A Japanese text treats this problem differently. The chapter begins with a discussion of equal angles. Then it states that the sum of the angles of a triangle is 180° and shows why. The students do not discover; they are told. (Even so, they do quite well on examinations that require originality.)

The American text achieves the worst of both worlds, depriving the pupils of the chance both to discover a mathematical fact and to see a mathematical proof—as contrasted to "experimental evidence."

In what I consider an ideal approach, the students would carry out the experiment I described and the text would not say anywhere what the sum turns out to be. After the class has lived with the problem, perhaps conjecturing that the sum is 180°, the teacher could ask, "Are you sure? Couldn't it be 179.45°?" After some discussion, the teacher should present the proof. It is up to the teacher to balance discovery and teacher telling.

(I fear that some teachers may read the *Standards'* advice "less teacher-telling" as "no teacher-telling.")

Under the heading "Critical Thinking" in a new California text was this challenge: "Is the circumference of a circle less than the perimeter of the square that surrounds it? Does that make sense? Explain."

Presumably it was inspired by the *Standards'* emphasis on thinking and communicating. However, I have no idea how to answer the second question nor what sort of explanation the pupil is supposed to give. The first question has the simple answer yes, which hardly develops skill in communication.

Elsewhere I found the assertion that the quotient of the circumference and the diameter cannot be written exactly as a decimal. This is utter nonsense. The quotient is an endless decimal that begins 3.14159, which we can write to as many places as we have time for. An endless decimal is perfectly respectable: witness, for instance, $\frac{1}{3} = 0.333333 \ldots$.

After browsing through hundreds of pages of several of the adopted texts, I had the uneasy feeling that they were put together in haste, with little or no class tests and no serious advice from mathematicians. The pages gave lip service to the letter and little service to the spirit of the *Standards.* Perhaps the constraints of time and money prevent publishers from doing the job right. On the other hand, they seem to have the time and money to clutter the texts with so many four-color illustrations that it was hard for me to find the mathematics. I had the uneasy feeling that the publishers added so many pictures, not to help the pupils learn, but to impress textbook-selection committees, who must wade through many competing candidates.

A report comparing texts in Japan and the United States concluded, "The Japanese books, averaging 200 pages, contained an average of seven chapters, each divided into two or three coherent sections, whereas U.S. textbooks, averaging 475 pages, contained an average of 12 chapters, each including about a dozen loosely related topics. . . . U.S. books devoted 19 percent of the space to irrelevant illustrations compared to 0 percent in Japanese books."

Separate research has shown that irrelevant illustrations do not improve learning. Clearly, the way American texts are produced is due for an overhaul.

Not only publishers, but teachers as well, will be responsible for bringing the *Standards* into the real world. Of particular importance are the teachers in elementary schools, for children often form a lasting impression of mathematics by the fifth or sixth grade. Moreover, because mathematics is a tight structure, where concept builds on concept and technique on technique, a pupil who misses a key idea may stay lost ever afterward.

Few teachers have the mathematical background to implement the *Standards*. The prospective elementary teachers I have had in class generally avoid science and mathematics classes, putting off the mathematics requirement to their senior year. Some arrive in class so weak that they cannot tell which is larger, $\frac{5}{6}$ or $\frac{1}{2}$. Nor can some of them show either arithmetically or pictorially that $\frac{2}{5} + \frac{1}{10}$ is equal to $\frac{1}{2}$.

One teacher, trying to put into practice what the *Standards* suggests, had her class figure out the cost of materials to build a park. To find the cost of the fence, they needed a formula for perimeter; unfortunately, she had given them instead the formula for area. Clearly, a teacher who is weak mathematically will soon be in hot water when conducting a class where anything can come up at any moment.

The *Standards* prescribes a far more extensive mathematical training of teachers than is customary today. For instance, teachers in grades 5 to 8 should know modular systems, matrices, trigonometry, coordinate geometry, geometry on the sphere, statistics, calculus, and the role of limits and infinity. I wish that the *Standards* had made the education of teachers its first priority, rather than propose a crash program that overhauls curriculum, pedagogy, assessment, and the preparation of teachers in one blow.

I fear that the *Standards* resembles an architect who has designed a beautiful bridge. Unfortunately, the firm that supplied the struts and beams makes them out of wood with a veneer of steel. The workers, trained as carpenters, not welders, put them together with nails.

The *Standards* will either bring us to the promised land or take us over yet another precipice. It is too early to predict. But one thing is clear. For the past century there has been essentially one long reform movement. As Sharon Dugdale, a mathematics educator and former teacher, advised me, "Its name changes more from decade to decade than its substance."

The suggestions I make in the next chapter are so simple and on such a small scale that I don't think of them as part of the perennial reform movement.

··· 13 ···

Some Proposals,
Modest and Immodest

The century-long effort to reform mathematics education should dis-
courage anyone from suggesting ways to improve the teaching and
learning of mathematics. Even so, being an optimist by nature, I will make
a few recommendations. Most of them have already been made more than
once by others but deserve to be repeated. In particular, I will make use of
The Learning Gap, by H. W. Stevenson and J. W. Stigler (1992). This little
book, based on a comparison of schools in China, Japan, and the United
States, should be read by teachers and parents.

Most of my suggestions are easy to implement. I will start with my
pet peeve.

··· Cartoonists ···

It doesn't help matters that cartoonists habitually make mathematics the
butt of their jokes, even if they only reflect society's unease with the subject. These cartoons are even reproduced in texts used in the preparation
of elementary teachers. Consider the messages a child gets when reading
these exchanges in the daily comics. The first is from *Peanuts:*

> "Say we cut an apple in half. We now have two halves."

> "That's fractions!! You're trying to teach me fractions! I'll never under-
> stand fractions! I'll go crazy!"

Isn't that a fine way to prepare a child for the study of fractions? Now how
about this in *Calvin and Hobbes?*

"I have a question about this math lesson."

"Yes?"

"Given that sooner or later we're all just going to die, what's the point of learning about integers?"

I admit that I found this amusing. However, why didn't the strip ask, "What's the point of learning to read or write or studying history?" (I am not suggesting that it should.) Why pick on mathematics?

I could quote many other examples from my collection of comics that reinforce the idea that it is okay for a child to fear or dislike mathematics. Never have I found one that presents mathematics in a favorable light. Given the critical role of mathematics in our society, couldn't a cartoonist try, just once, to do that? Following is a sample of what I mean, a poor offering but my own, intended to rankle cartoonists so that they will do better.

I have asked foreign visitors whether cartoonists in their countries poke fun of mathematics. "Certainly not," is their universal response, "mathematics is highly regarded and seen as necessary by the ordinary person." It may take a generation for our society to change its attitude. But cartoonists can accelerate the shift; at this moment they exacerbate the problem. They could be a part of the solution.

I shouldn't single out cartoonists. The Teen Talk Barbie doll was programmed to say, "Math class is tough." Why couldn't she have been programmed to say, "Math class is fun"? In most mathematics classes I have visited, the pupils were enjoying themselves. Luckily, there was such an outcry from the women's movement that Barbie stopped complaining. The manufacturer explained, "We never had any intention of discouraging girls or boys from studying math." No intention was needed. In our society, it comes without thought. It is about time that we think about the messages we are sending to our children.

The Learning Gap gave some attention to the role of these messages:

A nation that is falling behind in educating its children has reason to consider whether a process that can change the hairstyle of a vast majority of students might not be effective in motivating them to devote more attention to their studies. Those who pursue knowledge should be as worthy of emulation as sports heroes and rock stars. Developing models, however, requires a conscious effort. Bart Simpson, the "underachiever," reveals our national values: Americans view "nerds" and "bookworms" as models to avoid, but have no positive models of people whose accomplishments are based on knowledge, for students to imitate.

Advertisers could add to the standard celebrities some people whose contributions depend on a knowledge of mathematics.

••• Parents •••

A Midwest school grades parents, giving them credits for providing a quiet time for their children to do homework, reading with the children, and participating in school activities such as back-to-school night. This program reminds parents that they play a big role in their children's education, a role that is too often left entirely to the schools.

Reading is critical to success in school: in particular, in mathematics. Parents can help their children become good readers by reading to them. As the Commission on Reading reported in 1985, "The single most important activity for building the knowledge required for eventual success in reading is reading aloud to children." Jim Trelease's *Read-Aloud Handbook* gives the why's and how's and recommends titles.

The Learning Gap makes a big point of parents' involvement in school: "Rather than becoming disengaged from education once their children enter school, parents should preserve the high level of involvement they show during the preschool years. Unless they demonstrate that they value education and are willing to become involved in what goes on at school, their children will regard school as unconnected to the main events in their lives."

Teachers have told me that they often can predict a pupil's performance by noticing how much the parents are involved. As Barbara Gordon, a high school teacher in Liverpool, New York, said, "In my honors class, most of the parents show up at open house. In my other classes, very few."

No matter what attitude toward study prevails in society, a parent is in a position to counter it. This is made clear in the suggestion of John Wooton, director of player programs for the National Football League,

"By the time an outstanding athlete gets to college, he sees school only as an outlet for his athletic abilities. The only way to break this pattern is for the parents to take a greater role in their children's education." His advice applies to all parents, for all children face considerable distraction from their schoolwork.

A letter I received from Larry Askins, an Indiana high school teacher, complained about these distractions: "School musicals, band contests, tennis matches, track meets, the prom, football practice, owning a car (and paying for it with part-time jobs), are just a few of the activities that divert pupils from their homework."

Even parents with a limited mathematical background who cannot help their children with their homework can still remind them of its importance. My father did not go beyond the third grade, and my mother did not go beyond arithmetic. Though they never helped me directly, they assured me that my studies were important, telling me, "You have a job: to learn."

Even a single mother holding down two jobs can let her children know that she thinks school has a high priority. Even if she can see them only briefly in the course of a day, she can ask, "What did you learn today?" and remind them that mathematics not only is needed in many jobs but also is a tool for dealing with the decisions of daily life.

Provide your children with a well-lit place to study, preferably a desk, but the kitchen table will do. *The Learning Gap* reports, "The typical Japanese family occupies less than 900 square feet of space in what are sometimes derided as 'rabbit hutches.' Nevertheless, more than 80% of the families set aside space where their children can do their homework." Almost every fifth grader in their sample had a desk, while in the American sample, the figure was only about 60 percent.

I have heard parents say to their children, "I wasn't any good in math, so I don't expect you to be either." Wonderful psychology, and false. There are countless influences that may cause a person not to do well in a subject, none of which are transmitted by the genes. Instead, encourage your children to overcome any difficulties they may meet. After all, they are not required to invent brand new mathematics, only to learn a small part of what has already been around a long time.

Some teachers are reluctant to assign homework for fear of a backlash from parents. This has happened even in an algebra class, where practice is critical. Reassure the teachers that you expect them to assign what they believe is necessary and that you will support them.

Help your child set priorities so that first things come first. These days many children are so overscheduled that they don't have time to do their schoolwork. If that is true in your home, help them cut back. Don't be like the parent who explained to the teacher that his son didn't have time to do the homework because the son had taken up karate, gymnastics, and the piano.

When I was growing up in Los Angeles, except for homework, I led an easygoing life, playing baseball in the street, digging a fort in an empty lot, playing Ping-Pong at a friend's house. There's a good deal to say for free, unscheduled, spontaneous play—after the homework is done. One reads in *The Learning Gap* that "Chinese and Japanese children know that they will have free time only after they have completed the day's schoolwork. In America, leisure activities compete with schoolwork for the child's time."

As parents, you are in a position to enrich your children's mathematical experience. Have numbers play a natural part in your daily life. When you are cooking, have your child use the measuring cup. If you are putting up curtains or making a bookcase, have your child help measure. Ask for help in arithmetic when shopping; for instance, have your child compare the cost per ounce of the jumbo size against the giant. When traveling by car, ask your child to navigate, read the map, estimate the time of arrival at the next town, or figure out how much gas is left. There are similar questions when flying, including: "How high are we, in miles?" or "How many seats are there?" (a nice application of multiplication) or "What percentage are occupied?" "All told, how much did the passengers pay for this flight?" "How many gallons of fuel are used in a minute? In a second?" "How long does it take to go a mile?"

At the dinner table, bring up riddles or just simple drill questions. One of the games I played with my daughter Susanna, as early as when she was in the first grade, was "Find x." For instance, "If 3 times x is 15, what is x?" "If x times x is 6 more than x, what is x?" This gets children used to x as the usual name of an unknown quantity, well before taking algebra. There's no harm in introducing mathematics through games. For children, games are a serious affair.

The daily newspaper is full of numbers, with the sports pages crammed with decimals and percentages and, in the football season, with negative numbers as well.

Sustaining a daughter's interest in mathematics may require a strong maternal role. Mindy Bingham, author of the book *Things Will Be*

Different for My Daughter, emphasizes, "A mom's attitude has a lasting influence." She advises that even if a mother isn't comfortable with mathematics, "Never admit it. It's one time when you need to play act. Don't even joke about hating it."

One mother, whose mathematics is minimal, helps her daughter develop an ease with numbers by "pointing out everyday math when I use it—sewing, cooking, shopping. I involve the kids in my thinking."

Here are some other words of advice from specialists who want to rescue girls from the prejudice that "girls aren't good in math": Give your daughter an allowance and have her keep a budget and a log of expenses; expect as much in math from your daughter as from your son; check that her teacher doesn't, even unintentionally, favor the boys in class; introduce your daughter to the computer with "nonviolent, noncompetitive" software. (This last might be a good idea for sons as well.)

What about television? Though there are a few wonderful programs on television, the medium itself is the menace. It can hold a viewer under its spell for hours. The time spent in thrall, sitting in front of the screen is time not spent actively—playing, reading, helping.

If you can't ration television, just throw it out or, as the bumper sticker puts it, "Kill your TV." Such a ruthless act is possible, even in this day and age. We didn't have television when our children were growing up. Now, our daughter Rebecca doesn't have television in her home, and yet her children, ages 5 and 9, survive quite well.

However, throwing out the television may be too drastic. After all, there are some worthwhile programs on it. A few come to mind immediately, such as series on the Civil War, the Depression, and the civil rights movement, and art, nature, and science programs. Helping your children learn to choose which programs to watch may help them learn to think for themselves and make choices later in life. As adults, they will surely need personal discipline and moderation when they face the many temptations the world offers.

Clearly, parents have a major responsibility for the mathematics education of their children, even if they cannot help with the homework problems. Before we criticize the teacher, we should first check that we have done all we can to encourage our children. We should try to be the opposite of the parent who, when told by Barbara Gordon that her child was having trouble, replied, "You're the teacher. It's your problem, not mine."

••• Pupils •••

As children mature, they can take on increasing responsibility for their learning. However, some accept no responsibility, as if they were bystanders at their own education. The best example of this extreme is the student who sued the schools for "passing me from grade to grade and giving me a diploma, even though I couldn't read."

The opposite extreme is represented by the young Benjamin Franklin. When he was about 16 years old, he decided to improve his writing skills. In his *Autobiography,* he tells what he did to achieve this goal:

> I thought the writing in the *Spectator* excellent, and wished, if possible, to imitate it. With this view I took some of the papers, and, making short hints of the sentiment in each sentence, laid them by a few days, and then, without looking at the book, try'd to compleat the papers again, by expressing each hinted sentiment at length, and as fully as it had been expressed before, in any suitable words that should come to hand. Then I compared my *Spectator* with the original, discovered some of my faults, and corrected them. . . .

> I also sometimes jumbled my collections of hints into confusion, and after some weeks endeavored to reduce them into the best order, before I began to form the full sentences and compleat the paper. This was to teach me method in arrangement of thoughts. By comparing my work afterwards with the original, I discovered many faults and amended them; but I sometimes had the pleasure of fancying that, in certain particulars of small import, I had been lucky enough to improve the method or the language, and this encouraged me to think I might possibly in time come to be a tolerable English writer. My time for these exercises and for reading was at night, after work or before it began in the morning, or on Sundays.

Imagine a scale of self-reliance that stretches from the lad who sued the school at one end to Franklin at the other end. Each pupil fits somewhere in between and can choose to move toward the Franklin end. Such a choice may be necessary if there is turmoil in the classroom.

••• Businesses •••

Businesses spend over $25 billion a year on remedial education of their workers. I suggest that they divert some of that budget to improving mathematics education, perhaps by making funds available to bring mathematics-specialist teachers into the elementary grades. Or businesses could support visits by teachers to classes of all grades and to industry. Otherwise, teachers are cut off in the isolation of their own classroom. In the long run, such an investment may save far more than the present cost of remediation.

••• Mathematics Departments •••

Mathematics professors assume, perhaps subconsciously, that students majoring in mathematics will be going on to graduate programs in computer science or mathematics or to professional programs in law or medicine, or that they will become actuaries or systems analysts. It surprised my colleagues and me to discover that half of our graduates become high school teachers. At some colleges the number is as high as 70 percent. Each mathematics department should answer the question, "What is the best way we can help prospective teachers, math majors or not?" The answer will vary, but I doubt that it will be, "Exactly what we are doing now."

The *Standards for Teaching Mathematics* observes, "In the current culture of mathematics departments preservice teachers and those who work with them are treated as second-class citizens." Thomas Hungerford, a professor of mathematics, warned mathematicians in an article titled "Future Elementary Teachers: The Neglected Constituency,":

> Complaints about the mathematical preparation of incoming students are endemic in both college and high school mathematics departments. College professors tend to blame the high school teachers and they, in turn, blame the elementary school teachers for the sad state of affairs. Although this linear model of blame may be comforting to those at the top, a circular model (with blame for all) is a much better reflection of reality because elementary teachers are "trained" by the same college professors who complain about incoming students."

I urge my mathematical colleagues to read and act upon Hungerford's several specific suggestions.

J. R. C. Leitzel observed in *Undergraduate Mathematics Education* in 1991, "The mathematical preparation of elementary school teachers is perhaps the weakest link in our nation's entire system of mathematics education." However, mathematicians in Europe have a long tradition of being involved with precollege education. I mention, as just a few examples, A. N. Kolmogorov, Felix Klein, David Hilbert, Jacques Hadamard, and Stefan Banach, names familiar to all mathematicians.

The least we can do is develop a yearlong course that starts with an advanced view of arithmetic and reaches the fundamentals of calculus. This course would give future teachers at any level a perspective that would enrich their teaching and help them evaluate reforms and texts. Moreover, the material should have such substance and beauty that it could serve other students as well.

But not only what is taught is important. The way it is presented is just as crucial, for teachers will teach as they were taught. That means that instructors should not only lecture but also use, when appropriate, other approaches, such as cooperative learning and technology.

We already offer courses designed to serve business majors and engineers. It is about time that we meet the needs of future teachers.

This particular aspect of the education of future teachers brings up a more general problem.

••• The Professional Preparation of Teachers •••

In many states, prospective teachers enter a teaching program and a year later receive a credential that certifies its owners are "experts" in the sense that they can teach more effectively than mere laypeople who lack a credential. However, numerous experiments show that laypeople are just as effective in the classroom as certified teachers.

In their book *Professional Teaching Expertise: Fact or Myth,* L. Jennings, S. George, and A. Schell conclude, "At least in the case of elementary school teachers there is little or no relation between the amount of professional preparation and teaching effectiveness. However, research reveals appreciable differences in teachers' effectiveness, but these differences appear unrelated to preparation. These contrasts seem to result from naturally occurring differences."

Their evidence is overwhelming. In part it is based on controlled experiments in which certified teachers and laypeople teach the same lessons. It is also based on the performance of people hired because there were not enough certified teachers and people given credentials based on their mastery of subject matter. Consequently, the authors recommend, "The principal requirement for eligibility to teach in elementary school should be the ability to produce unusual growth in pupils' achievement."

That means that the credential would be awarded on the basis of performance, not on the basis of the number of classes taken. "No other requirement significantly predicts success." Such a criterion would bring teaching into line with those professions, such as the practice of medicine, that involve carefully monitored apprenticeships.

••• Schools •••

Finally, I come to the schools. *The Learning Gap* has several suggestions for bringing our schools up to the caliber of, say, those in Japan. Its most important recommendation is, "Decrease the teaching load of elementary school teachers. Until teachers have *adequate time* to prepare lessons, work outside of class with individual students, and perfect their teaching practices by interacting with each other and with master teachers, it is going to be difficult, if not impossible, to change what children learn in school."

Such a change may require increased taxes for the support of our schools. Unfortunately, *tax* has become a conversation stopper even though we are not overtaxed in comparison with other industrialized countries. Our schools are the last place to economize. Short-term savings there will show up as long-term expenses later. That 85 percent of the 1.5 million people residing in our jails and prisons did not graduate high school should inspire some reflection. A strong school system means a more sophisticated workforce that will, in turn, pay taxes and be less likely to need welfare.

It's odd that a nation that can pay at least $200 for each of the 70,000 seats at Superbowl XXIX and come up with $16 million for the 16,000 seats at the Tyson-McNeeley boxing match and $65 million to watch it on television (1.52 million sets at an average fee of $43) lacks the resources to support its schools. Maybe we should ask ourselves, "What is important?" If we don't, then, like the mathematics professors who only complain, we should stop criticizing our schools.

••• Accept the Split Personality ••• of Mathematics

My final suggestion may be the only original one in this chapter. Yet it is so obvious that I am sure I am not the first to make it.

The battle between basics and understanding, between routine calculation and logical thinking, has persisted for a century in mathematical education. More than one reform has promised to resolve the conflict and failed.

There may be only one way to settle the issue once and for all. Instead of continuing the age-old battle, exploit the dual nature of mathematics. There should be *one course devoted to calculation* and *one course devoted to concepts and solving problems.* There is meat enough for both.

The calculation course, besides offering practice in arithmetic, would include the abacus, slide rule, calculator, computer, and mental arithmetic. The concept course would develop the underlying ideas, such as the notion of a fraction and the decimal system of notation, the history of such ideas, and the solving of nonroutine problems, as well as skill in written and oral communication of mathematics. With such a division, neither the basics nor the concepts can drown the other out.

If there aren't enough teachers qualified to conduct the "concepts" course, then hire some of the engineers and scientists stranded at the end of the Cold War. They may have to be put on a higher pay scale, which will irritate other teachers, who will have to adjust. On my campus, professors of engineering, law, medicine, and economics have higher salaries than the other faculty, but the rest of us have learned to live with the inequity.

Putting this suggestion into practice may take time. But, by contrast, it shows that the others are comparatively simple and could be implemented tomorrow. There is no need to wait and see how the latest effort in the reform movement turns out. In the meantime, there is a good deal each of us can do.

PART TWO

From High School to Kindergarten

··· 14 ···

How to Read
Mathematics

I usually read a newspaper backward, starting with the sports section, then the weather, then Ann Landers, and finally the front section. I even read the front section backward, since the most important news tends to be dull and is put in the back pages. That is no way to read mathematics. Even reading newspapers, magazines, and novels in the direction intended by the writers does not prepare one to read mathematics or, for that matter, any text written in the concise language of mathematics, such as an electronics manual or a physics book.

To follow the steps in a line of reasoning, the reader must become an active participant, practically a co-author. It is not just a matter of reading more slowly than usual or of going over the same page several times. It's a question of being alert and suspicious, letting nothing get by. Nothing can be skipped. The reader should work as hard as the author.

There's another way of looking at the challenge of reading mathematics. The author tries to choose the words that will make the logic clear to readers of many different thinking styles. The reader, on the other hand, must deal with only one person, the author. Both have to play active roles. In this sense again, *mathematics* is a verb, not just a noun.

In short, the reader should make sure that each step is clear: stop to draw a missing diagram, check a calculation, or make up examples to illustrate some claim made by the author. The experience in daily life that comes closest to this is not reading a newspaper. Rather it is like two people playing foot bag, also known by the trade name Hacky Sack. The object is to keep a stuffed bag, about the size of a tennis ball, in motion without touching it with the hands.

I agree with the advice that the columnist Marilyn vos Savant gave a reader who had written, "I simply cannot get algebra through my head. Do you have any advice?" She replied, "Buy the most elementary high school algebra book and work through it from the beginning. Reason through the problems in addition to using the formulas. . . .

"Why do so many people have difficulty with mathematics? Because the study of mathematics requires perfection. Every step of the way strict attention to the smallest detail is absolutely necessary."

But this is true not only in reading mathematics. It applies to carpentry, tennis, singing—almost any activity that can be done well or poorly.

There's another contrast between a newspaper and mathematics. Mathematics is written in a terse, concise style. In a newspaper, there's a certain looseness or redundancy. You can delete words and still be able to understand a sentence. As an example, the last three sentences can be shortened to, "Mathematics is terse. A newspaper is redundant. You can delete words and still understand a sentence."

It is even possible to delete letters from words in ordinary writing and still get the drift. If I say that I had bkfst, lnch, and dnr today, the reader would know I didn't go hungry. Similarly, the personal license plate "TRTH SKR" is easy to decipher. (It's "truth seeker" without its vowels.) But in mathematics, each symbol counts. The following five statements all say the same thing. The first is ordinary prose. The fifth is mathematics. The middle three statements show a gradual transition from the easygoing, long-winded first one to the businesslike, terse final one.

Three is the positive number which, when you multiply it by itself, gives nine.

Three is the positive number whose square is nine.

3 is the positive number whose square is 9.

3 is the positive square root of 9.

$3 = \sqrt{9}$.

Compare the fifth with the first sentence. In the last one, there isn't a symbol to spare. (By the way, the square root symbol $\sqrt{}$ probably grew out of a dot. A few hundred years ago, the square root of 9 would have been written .9. Some people feel that the symbol is a misshapen r, for "root," but there is no evidence for this belief.) For practice, note that

$\sqrt{36} = 6$ and $\sqrt{49} = 7$. In those two cases, the square roots are convenient numbers, but a square root need not be so nice. For instance, the square root of 40 lies between 6 and 7. I invite the reader to estimate $\sqrt{40}$ to a few decimal places either with experiments by hand or with a calculator.

Just to make sure my point is clear, I will give another example, starting with ordinary prose, then finishing in the succinct language of mathematics.

> When you multiply a number that is between 0 and 1 by itself, you get a number that is smaller than that number.

> The square of a number that is between 0 and 1 is smaller than that number.

> If r is a number between 0 and 1, then r^2 is less than r.

> If $0 < r < 1$, then $r^2 < r$.
> (The last version uses the symbol $<$, which is short for "is less than.")

The reader who skips over any symbol in the fourth, bare-bones sentence will miss its drift. It does not lend itself to speed reading. Instead, it invites slow reading. Though much shorter than the first statement, it says the same thing. The reader should check that it makes sense by testing it for some number between 0 and 1, say, $r = 0.7$. The square of 0.7 is 0.49, which is indeed smaller than 0.7. Doing such a check provides an extra cue to the memory and a "speed bump" to slow the reader to the pace appropriate for reading the language of mathematics. So does saying every symbol to yourself or aloud.

In Chapter 15 I use the term *whole number*. Just to be sure that we all agree on what it means, I will stop to define it. A *whole number* is any of the numbers 1, 2, 3, 4, 5, . . . that we use when counting. Some people call them the *natural numbers* and even include 0 as well, for "in the beginning was the void."

An *integer* is more general than a whole number. It is either a whole number, 0, or the negative of a whole number. In the diagram on the next page, which shows the number line, the integers are shown by regularly spaced marks. The whole numbers are the integers that lie to the right of 0, the positive integers. The negative integers lie to the left of 0. Zero itself is neither positive nor negative.

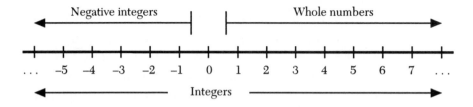

Following are short definitions of a few words and symbols that we will be using. Later we will add to the list.

Word or Symbol	Meaning
<	less than
>	greater than
=	equal
≈	approximately equal
$\sqrt{}$	square root
whole number	1, 2, 3, 4, 5, ...
integer	..., −4, −3, −2, −1, 0, 1, 2, 3, ...

Mathematics is not the only writing that must be read word by word, symbol by symbol. Other examples are the warning on a pesticide, a recipe for minestrone, and even an electronics manual. We must switch gears to slow to get the essential details. Reading the daily newspaper does not develop the right pace, the necessary patience and care.

One more suggestion. Pay particular attention to the definitions. Copying them will focus your attention and help you remember them.

I remember a graduate student in mathematics who said, "I'm going to skip the details when I read and concentrate on the ideas. Later I'll go back to fill in the steps." Two years later he told me that he had learned nothing. To keep a balance between the nitty-gritty and the big idea, the reader should concentrate first on the details, then put the pieces together.

To read mathematics, go slow, read each symbol and word, and check each statement. The second time through, put it all together. Even if you're not used to this kind of reading, you'll find that with practice it will become quite natural. Keep these words in mind:

After all these years
Grandma's English still is poor,
her accent thick.
She understands, she is understood.
It's good enough, she gets by.

I'm like her:
I can replace a light bulb,
boil an egg,
tie my shoes.
I stop there.
Why not wire a house,
bake bread,
learn that knot
that cinches over a tarp?

Why did I settle for this niche,
selling myself short?

How high could I jump?
How well throw a pot
or paint a water color?

Had I grown up in China
I could speak Chinese.

Chapter 15 gives you a chance to practice reading mathematics (not Chinese).

··· 15 ···

You Will Never See a Large Number

One day a reporter from the *Sacramento Bee* called me and said, "There's an $8 billion deficit in the California budget. Is there some way to help readers feel what 8 billion really is?"

"Well, that's like 400,000 workers, each earning $20,000."

"That's too abstract."

"Then how about this: How much of San Francisco could you cover with 8 billion dollar bills?"

"That would be more concrete."

"I'll work it out and phone you back."

I knew that San Francisco is roughly a square, with each side 7 miles long, so it has an area of about $7 \times 7 = 49$ square miles. Next I measured a dollar bill, which turned out to be $6\frac{1}{4}$ by $2\frac{5}{8}$ inches. After turning these numbers into decimals, 6.25 and 2.625, I ground away on my calculator to discover that it would take about 12 billion dollar bills to cover the city. So 8 billion would cover only about two thirds of San Francisco. But it was more than enough to cover the entire island of Manhattan or, more appropriately, Sacramento itself. While we're at it, it might be useful to note that a national debt of $5 trillion would cover all of Vermont and New Hampshire, with enough left over to take care of New York City, Chicago, Los Angeles, and a few other cities of your choice.

So a billion or a trillion, numbers that we see in the daily paper, appear to be gigantic. They are beyond our comprehension, even though we see them whenever we read about the buyout of a corporation or the cost of an earthquake or a submarine.

I like to think of $1 billion as "about 32,000 workers, each earning

$32,000." That brings it down to a human scale but still keeps it respectably large. In this approach, $1 million is just 1,000 workers, each earning $1,000.

In any case, a billion (1,000,000,000) and a trillion (1,000,000,000,000) have so many zeros that it's hard to read them. It's easier to tell how many times you have to multiply 10 by itself to get such a number. A billion, which has nine zeros, is the product of nine 10s, $10 \times 10 \times 10 \times 10 \times 10 \times 10 \times 10 \times 10 \times 10$, which is written 10^9 (read as "ten to the ninth"). Since a trillion has 12 zeros, it is written 10^{12}. This so-called exponential notation has the advantage of brevity but the disadvantage of concealing the true size of the number, since we no longer see a long string of zeros.

While I was writing this chapter, my 6-year-old granddaughter, Helaina, called and asked, "What is a googol?" It turns out to be the next to the largest number defined in a dictionary. It is 1 followed by a hundred zeros. It first appeared in 1940 in *Mathematics and the Imagination* by Edward Kasner and James Newman. The name is due to Kasner's 9-year-old nephew. So is the name "googolplex," which is 1 followed by a googol of zeros, the largest number in a dictionary. Gigantic though it may seem, mathematicians have run into much larger numbers in their study of properties of the whole numbers.

Back in the third century B.C., Archimedes wrote on the question of whether a number is so large that it is infinite:

> There are some, King Gelon, who think that the number of grains of sand is infinite; and I mean by the sand not only that which exists about Syracuse and the rest of Sicily but also that which is found in every region whether inhabited or uninhabited. . . . But I will try to show you that of the numbers named by me some exceed not only the number of grains of sand that would fill the earth, but even the entire universe.

Assuming that a grain of sand is the size of a poppy seed and that the diameter of the universe is at most ten thousand times as large as the diameter of the earth, Archimedes concluded that no more than 10^{51} grains of sand could fit into the universe.

I'm sure we all agree that a million, a billion, and a trillion are big numbers and that 1 and 10 are small numbers. We may not all agree about 100 and 1,000. This raises the amusing question, "Where do the small numbers end and the big numbers begin?" I leave this for the reader to ponder.

It turns out, however, that in mathematics even a trillion is small potatoes. It really is a small number. I will go so far as to say that no one has ever seen a really large number and never will. Of course, I had better justify such a claim, especially after noting that $1 trillion could cover two states.

To show what I mean, I will describe a simple game of solitaire that can be played with paper, pencil, and the whole numbers 1, 2, 3, 4, ... As I explain at the end of the chapter, this game has serious implications in mathematics.

Take a whole number. If it has exactly two distinct factors, itself and 1, it is called a *prime.* For instance, the first few primes are 2, 3, 5, 7, 11, 13, 17, 19, and 23. Each whole number from 2 on is either a prime or a product of primes in exactly one way. (Note that 1 is not a prime, for it has only one factor, itself.)

Now, any whole number from 2 on is either a prime, the product of an even number of primes, or the product of an odd number of primes. For instance,

$$15 = 3 \times 5$$

is the product of two primes, an even number of primes. Let us call a whole number that is the product of an even number of primes an *Evener.* So 15 is an Evener. However,

$$30 = 2 \times 3 \times 5$$

is the product of three primes, an odd number of primes. Such a whole number we call an *Odder.* We will also consider every prime number to be an Odder, for we may view it as the "product" of just one prime, and 1 is odd. For instance, 11 and 29 are Odders. Call the number 1 an Evener (because it involves zero primes, and 0 is even).

Now break the whole numbers into two teams, the Odders and the Eveners. The following chart shows which of the numbers from 1 to 15 are Odders and which are Eveners.

Number	1	2	3	4	5	6	7
Factorization	–	prime	prime	2×2	prime	2×3	prime
Verdict	Evener	Odder	Odder	Evener	Odder	Evener	Odder

Number	8	9	10	11	12	13	14	15
Factorization	$2 \times 2 \times 2$	3×3	2×5	prime	$2 \times 2 \times 3$	prime	2×7	3×5
Verdict	Odder	Evener	Evener	Odder	Odder	Odder	Evener	Evener

Up to 15 there are 8 Odders and only 7 Eveners. Imagining a contest between the Odders and the Eveners, we can say that the Odders are ahead, leading by a score of 8 to 7. At the start, Eveners led briefly 1 to 0, since 1 is an Evener. But then the Odders catch up and by 3 are ahead 2 to 1.

Then comes $4 = 2 \times 2$, an Evener, and the score changes to 2 to 2, a tie. Then, with 5, the Odders pull ahead and the score is 3 to 2. But along comes $6 = 2 \times 3$, an Evener, and the Eveners catch up again. The score is then 3 for Odders, 3 for Eveners. But then comes 7, and the Odders pull ahead, leading 4 to 3.

This raises a question: Do the Eveners ever lead the Odders, other than briefly at the start? G. Polya (1887–1985) raised this in 1919 and checked that the Eveners don't take the lead at least through the number 1,500. R. Lehman checked that the Eveners don't take the lead up through 1,000,000. That fact might persuade us that the Eveners never take the lead. After all, a million seems like a pretty big number.

Surprise. Lehman showed in 1960 that if you keep the game going up through the number 906,180,359, which is almost a billion, the Eveners lead the Odders by 1 at that point. In 1980, M. Tanaka showed that the first time this happens is at 906,150,257.

So even running a test up to a million may suggest the wrong conclusion. When you realize that the whole numbers go on and on forever, the first million or billion or even trillion of them are really a very tiny sample. They form only the visible front end of the endless line of whole numbers.

It is safe to say that any whole number we will ever see, any number that we could write down, even with a string of zeros a mile long, is small. The truth about whole numbers lies in the realm of numbers "way out there," well beyond the limits of our experiments. Even computers that execute a billion operations a second will make only a small section of the whole numbers visible to us.

But let's play another game of solitaire, a variant of the first one, invented by F. Mertens (1840–1927) in 1897. It provides an even more dramatic warning to be wary of small numbers.

Some numbers are a product of primes with no prime appearing more than once. Another way to describe these numbers is to say that "they are not divisible by any square number larger than 1." Four examples of such numbers are 2, 15, 165, and 858, since

$$2 = 2, \quad 15 = 3 \times 5, \quad 165 = 3 \times 5 \times 11, \quad \text{and} \quad 858 = 2 \times 3 \times 11 \times 13.$$

Observe that every prime number is of this type.

In other numbers at least one prime in the factorization is repeated. The numbers 4, 45, and 27 are examples of this, since

$$4 = 2 \times 2, \quad 45 = 3 \times 3 \times 5, \quad \text{and} \quad 27 = 3 \times 3 \times 3.$$

We will pay no attention to numbers of this second type. Let us throw them away, so to speak. We will be looking only at the whole numbers of the first type, where each prime in the factorization appears exactly once and no more. Let's call these numbers *S-numbers,* with S standing for "Special" or "Square-free." We will include 1 as a special case. Here is the list of S-numbers up through 30 with their factorizations:

S-Number	1	2	3	5	6	7	10	11	13
Factorization	–	prime	prime	prime	2×3	prime	2×5	prime	prime

S-Number	14	15	17	19	21	22	23	26	29	30
Factorization	2×7	3×5	prime	prime	3×7	2×11	prime	2×13	prime	$2 \times 3 \times$

It's a good idea to check this yourself and then go beyond, in order to develop a feeling for S-numbers.

An S-number is either an Odder or an Evener. We will call them S-Odders and S-Eveners, respectively. This game involves these two types of numbers, but it is quite different from our first game.

This time we pick a whole number, which we will call *n,* for the lack of a better name. Then we see how many S-Odders there are up through *n* and how many S-Eveners there are up through *n.* Finally, we find the *difference* between these two counts.

To make sure that this is clear, let's look at the case when the number *n* is 30. Up through 30 the S-Odders are

$$2, 3, 5, 7, 11, 13, 17, 19, 23, 29, \text{ and } 30.$$

So there are 11 S-Odders. The S-Eveners that are not larger than 30 are

$$1, 6, 10, 14, 15, 21, 22, \text{ and } 26.$$

There are 8 S-Eveners less than or equal to 30.

Since there are 11 S-Odders and 8 S-Eveners, the difference in this case is 3, which isn't very large. Offhand, we would expect the number of S-Odders to be roughly equal to the number of S-Eveners, for any number *n.* In other words, we would expect the difference between the number of S-Odders and the number of S-Eveners not to be very large.

Mertens compared their difference to the square root of n, $\sqrt{n}$. (Later in this chapter I will indicate why $\sqrt{n}$ is important here.) For instance, when n is 30, the square root of n is about 5.5, as you may check on your calculator. So in the case of the S-numbers up through 30, the number of S-Odders differs from the number of S-Eveners by less than the square root of 30, since 3 is less than $\sqrt{30}$.

It doesn't take long to see what happens when you extend the list up through, say, 65. Take a few minutes to do the arithmetic. You should find 20 S-Odders and 20 S-Eveners that do not exceed 65. In this case, the difference is 0, and it is less than the square root of 65, which is about 8.1. So again the difference between the numbers of the two types, the S-Odders and the S-Eveners, is less than the square root of the number we are going up to, up to 30 in the first case, up to 65 in the second case. If you go up to, say, 100, you should find 30 S-Odders and 31 S-Eveners, for a difference of 1, which we note is indeed smaller than the square root of 100, which is 10.

After making similar calculations all the way to 10,000, Mertens concluded that it is "very probable" that the difference will always be less than the square root for all whole numbers. (He was not interested in which type was ahead; it turns out that each of the two types takes the lead infinitely often.)

This conclusion, known as the *Mertens conjecture,* inspired a great deal of study throughout this century, for an important reason that I mention later. I'll introduce a shorthand in order to describe what was discovered.

First you pick a whole number n. (We picked 30 first and then 65 and 100.) Then you find the difference between the number of S-Odders and the number of S-Eveners that are not larger than n. Call this difference simply $D(n)$, which we read as "D of n." As we saw, $D(30) = 3$ and $D(65) = 0$, and as we mentioned, $D(100) = 1$. We will use the symbol $\leq$ to mean "does not exceed" or, in other words, "is less than or equal to." In our shorthand the Mertens conjecture reads

$$D(n) \leq \sqrt{n}$$

for every whole number. In this game the race is between $D(n)$ and $\sqrt{n}$, not between the S-Eveners and the S-Odders.

During the period from 1897 to 1913, L. von Sterneck checked that the Mertens conjecture is true all the way to 5,000,000. He observed something even stronger than that conjecture. For every n beyond 200 but not exceeding 5,000,000, $D(n)$ was even less than half the square root of n,

$$D(n) \leq \frac{\sqrt{n}}{2}.$$

Or, to put it in decimal notation, $D(n) \leq 0.5\sqrt{n}$. So it is natural to conjecture that this might hold for all n beyond 200. However, as you might now expect, 5,000,000 is such a small number that von Sterneck's conjecture could be wrong.

Henri Cohen and François Dress in 1979 computed $D(n)$ for all n up through 7,800,000,000 and found that $D(n)$ is *larger* than half the square root of n when n is 7,725,038,629. So von Sterneck's conjecture was wrong. By the way, that is the first time, for n larger than 200, that $D(n)$ exceeds $0.5\sqrt{n}$. However, they found no example that violated the Mertens conjecture. In fact, $D(n)$ is no larger than $0.6\sqrt{n}$ for the numbers that they looked at.

But that is not the end of the story. Even though the Mertens conjecture was verified for all whole numbers up through 7,725,038,628, which is about 8 billion, it turns out to be false eventually.

In 1984, A. M. Odlyzko and H. J. te Riele, combining extensive theory with extensive computing, showed that there are an infinite number of numbers n for which $D(n)$ is greater than $1.06\sqrt{n}$. If you would like to see a specific number where the Mertens conjecture fails, I'm sorry, but you will be disappointed. Odlyzko and te Riele say, "Our proof is indirect, and does not produce a single value of n for which $D(n)$ is greater than $\sqrt{n}$. In fact, we suspect that there are no counter-examples to Mertens conjecture less than 10^{20} or perhaps even 10^{30}." If their suspicion is correct, even the fastest computer might not find the first counterexample in our lifetime.

But why are the two games we played more than games? Because had Polya's conjecture or the Mertens conjecture been true, then the Riemann hypothesis would have been confirmed.

The Riemann hypothesis asserts that a certain infinite set of special points in the plane, defined with the aid of calculus and complex numbers, lies completely on a certain line. Though Riemann proposed it in 1859, it is still not known whether it is true or false. So far, it has been checked for 1.5 billion points, but, as we have seen, 1.5 billion is a rather small number. Moreover, in 1914, G. H. Hardy proved that an infinite number of the special points do lie on a line.

The Riemann hypothesis is important because it implies many other results in number theory. For instance, if true, it would even give us information about the size of $D(n)$ when n is large. Mertens conjectured that

$D(n)$ is less than the square root of n, $\sqrt{n}$. In other words, he conjectured that the quotient, $D(n)/\sqrt{n}$, stays less than 1. The Riemann hypothesis implies that the quotient grows very slowly, more slowly than any power of n, even more slowly than $n^{1/100}$, for example ($n^{1/100}$ is the positive number which, when multiplied by itself 100 times, yields n). The reverse holds as well. That is, if $D(n)/\sqrt{n}$, does grow so slowly, then the Riemann Hypothesis is true. So our game with the S-Eveners and the S-Odders is a serious matter indeed. However, no one knows how the quotient $D(n)/\sqrt{n}$, behaves as n increases.

That hypothesis still challenges us as the most important problem in all of mathematics. It stands, a lofty mountain that no one has scaled, though many have struggled to reach its summit.

You may be thinking, "I can see that this would interest mathematicians. I even find it fascinating. But what is the importance of this to the world outside of mathematics?" An example in Chapter 8 answers this question. There I described a bit of number theory concerning the product of two large primes, which mathematicians never thought would be of practical use. Lo and behold, it turned out to be the basis of a code used to transmit confidential information.

The fate of Polya's and Mertens's conjectures warns us to view a million, a billion, or even a trillion as "small" numbers. Those conjectures were of interest in their own right and because of their connection to the Riemann hypothesis. But it turns out that a very important practical concern should also persuade us that some numbers we usually think of as "large" may not be so large after all.

In 1977, Rivest, Shamir, and Adleman invented a coding system now widely used in banking, in the armed forces, and in nuclear power plants. (I described the basis for this system in Chapter 8.) The security of this "RSA code" depends on the fact that it takes a long time to factor a "large" number. To break the code, an intruder would have to find the prime factors of a certain number that was not secret.

The three inventors published a 129-digit number, dubbed RSA 129, predicting that with the methods then available it would take 4 times 10^{16} years to factor it—by which time the sun would no longer be shining. They also predicted that even with improved technology, no one could break the code until some time in the twenty-first century. Yet RSA 129 was cracked in 1994.

In the 17 years since RSA 129 was proposed, mathematicians invented faster ways to factor numbers and computer scientists developed faster computers. It became possible for 600 Internet volunteers, working for eight months at their computers, to carry out the 10^{14} calculations needed to factor RSA 129. So perhaps we should view even a number 129 digits long as "small," since a code based on such a number can be cracked. To play it safe, RSA codes typically are based on numbers 135 to 150 digits long. Soon even these numbers may be too small to guarantee the security of the codes.

Now let's look back at the two conjectures. Polya's conjecture about the race between the Eveners and the Odders turned out to be false at 906,150,257, though true up through 1,000,000. The Mertens conjecture, concerning the difference between the number of S-Eveners and the number of S-Odders, though true up through 7,725,038,628, is not true in general. This should make us hesitate before offering guesses based on a mere few million or few billion cases. Beyond that comfortable world of "small numbers" we can deal with by paper and pencil, or calculator, or even computer, lies an endless jungle of whole numbers, a realm that holds secrets it may never reveal. Exploration in that remote land must be carried on with the aid of deep mathematical theories. Still, some parts may remain forever beyond our reach. We may never meet all the strange plants and animals that inhabit that exotic realm.

··· 16 ···

The Car and
Two Goats

A few years ago in one of her columns, Marilyn vos Savant posed this brainteaser, submitted by a reader:

> Suppose you're on a game show, and you're given a choice of three doors. Behind one door is a car; behind the others, goats. You pick a door—say Number 1—and the host, who knows what's behind each door, opens another door—say, Number 3—to reveal a goat. He then says to you, "Do you want to pick door Number 2?" Is it to your advantage to switch your choice?

Her answer and further explanations in later columns called forth an avalanche of chastising letters from Ph.D's across the nation—professors at reputable institutions. I quote a few, omitting the writers' names, because they have already been embarrassed enough. They were wrong and Marilyn was right.

> I'm very concerned with the general public's lack of mathematical skills. Please help by confessing your error. . . .

> I am in shock that after being corrected by at least three mathematicians, you still do not see your mistake.

> Maybe women look at math problems differently than men.

I will resist the temptation to give the answer because my goal in this chapter is to convince you that you can think mathematically and even more clearly than some mathematicians. I will only suggest a way to approach this problem, a way that would apply to many problems. I am sure that, with just a little guidance, you will unravel the brainteaser.

First of all, go slow. Think about the question awhile, so that it becomes absolutely clear. Perhaps you develop an opinion on whether the game show contestant should stay or switch, or maybe it doesn't matter. But before you become wedded to your opinion, carry out the following experiment. It will help you develop a deeper insight into the question.

Get hold of three identical tin cans, paper or Styrofoam cups, or three 35-millimeter film canisters, or any three identical, opaque containers. I'll phrase the experiment in terms of the canisters, since they are the easiest to find.

Put a small piece of paper in one of the canisters (or anything that won't make a noise as the canisters are moved). The three canisters stand for the three doors. The piece of paper stands for the car, and the empty canisters represent the doors that conceal the goats. The following figure shows the tools of the experiment.

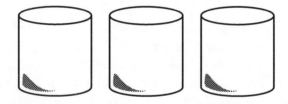

You pretend to be the host and a friend can play the role of the contestant. (If your friends are too busy, you can play both roles. Just don't peek into the canisters until it is time.)

Shuffle the canisters, not letting your friend look inside any of them. Then have the friend choose a canister, hoping to get the one that has the wad of paper. Look inside the other two canisters and show your friend an empty one. However, *insist* that your friend switch from the initial choice. Record whether the switch was a hit or a miss. Do this whole experiment 50 times, recording hits and misses. A table like the one shown here would help.

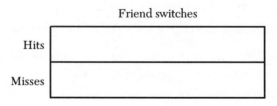

Friend switches

Hits	
Misses	

Although 50 experiments seem like a lot, they go quite quickly. Then total up the number of hits and the number of misses.

Next do the same experiment 50 times, but now have the friend not switch from the initial choice. Fill out a table like the following.

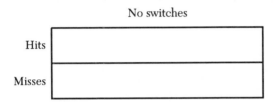

After completing these 100 experiments, what do you think? Do both tactics yield a similar number of hits? Do you think it is wiser to switch or to stay put, or doesn't it matter? Do the data you collected suggest a way to analyze the problem?

It is best that your friend doesn't try to solve the problem. If both of you work on it, the mood may change from a rational calm to a competitive excitement, which can interfere with your concentration.

If, after thinking some more about the question, you still are not sure about the answer and are not ready to explain it, then do the following. (Keep in mind that just citing experimental data is not an explanation. The data may convince you that something is true, but they do not explain it.)

Get one more canister and perform a similar experiment, using four canisters instead of three. Put a wad of paper in one canister. After your friend chooses a canister, look in the remaining three and show the friend two empty canisters. The friend then faces a choice between the two other canisters. Carry out the same experiments as before. Think over the results you get. What do they suggest? Do you see a way to explain what happens?

Performing these experiments not only gives you some clues, it also slows you down from the common frenzy of everyday life, so you can focus on just one thing for a period of time.

If you still do not see how to explain what is going on, then use ten canisters. Put the wad in one of them. After your friend chooses a canister, look in the other nine. Show your friend eight empty canisters out of those nine and remove all eight. Again that leaves just two canisters. Conduct a similar experiment.

I am confident that you will solve this problem, so confident that I do not include the answer anywhere in the book, not even in fine print upside down hidden in the back matter. You will probably, along the way, calculate the fraction of times that switching will pick the car and the fraction of times that not switching will pick the car. Using these fractions, you will be able to explain the brainteaser completely. Then you will have to admit that you can think mathematically. You just needed the opportunity.

··· 17 ···

Five Things
You Can Do with
Two Numbers

One day when I was playing arithmetic with my grandson Jason, I asked him, "What is 3 times 5?" After some thought he replied, "15." "Correct. Now what is 5 times 3?" He started all over from scratch, finally coming up with 15 again. He did not realize that five threes is the same as three fives. After all, the fundamental properties of numbers are not coded in our genes. For this reason, it may be worthwhile to review the basics of our number system. Since you can do only five fundamental things with two numbers, there are just a few basic principles.

First of all, let us find out why 5×3 is the same as 3×5. We can explain it by the way multiplication of whole numbers is introduced in grade school. "Five times three" means forming five groups of three things, as in Figure 1.

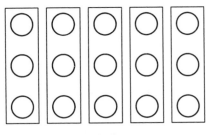

Figure 1

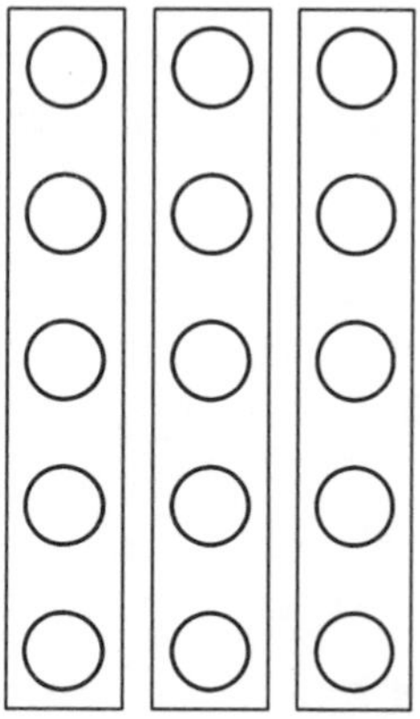

Figure 2

"Three times five" asks us to handle the numbers differently: to form three groups of five things, as in Figure 2. Why should my grandson see any connection between them? Clearly, they are quite different.

You can see a connection between Figures 1 and 2 when you group the dots in Figure 2 as shown in Figure 3.

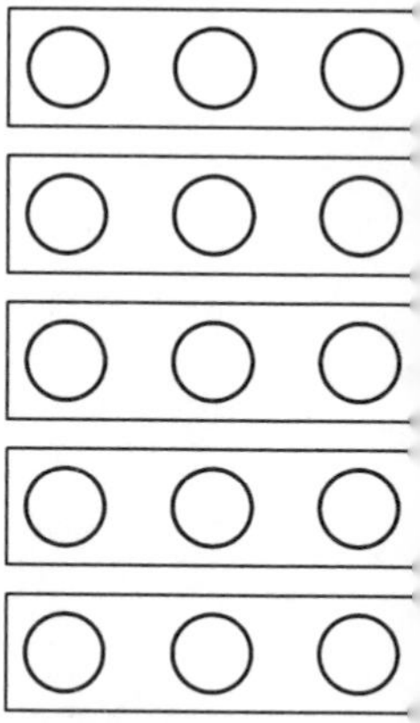

Figure 3

Comparing Figures 2 and 3 shows you why 3×5 equals 5×3. A carton holding a dozen eggs also shows why you can switch the order of two numbers without changing their product. We can think of the dozen eggs as two groups of six eggs each, or as six groups of two eggs each, as in Figure 4.

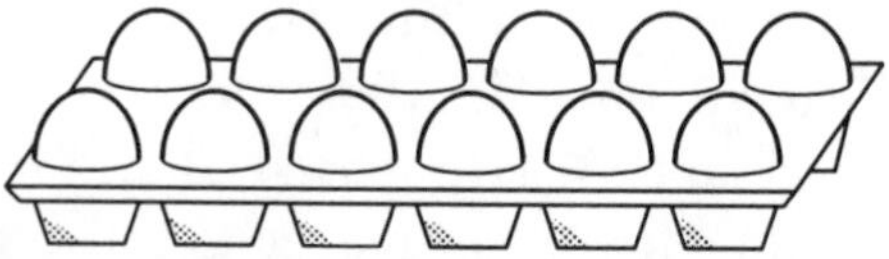

Figure 4

In a way, the egg carton is an optical illusion, because we can view it in two different ways.

Why then is $3/5$ times $2/3$ the same as $2/3$ times $3/5$? No egg carton can explain that equality. However, if you think of multiplication as finding the area of the rectangle in Figure 5, the answer is easy.

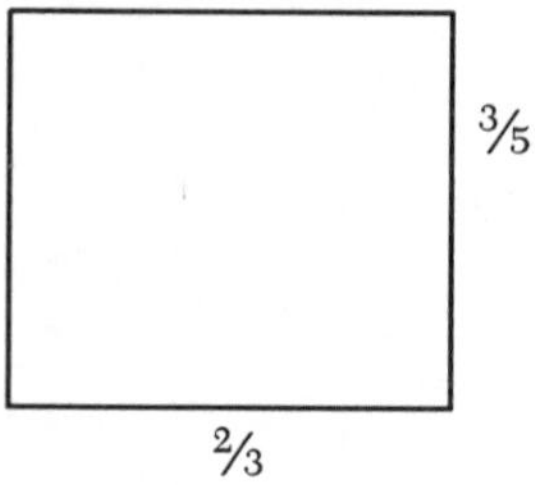

Figure 5

That area is the product of the two dimensions of the rectangle; order doesn't matter. That the product of two numbers does not depend on the order in which they are written is called the *commutative property* of multiplication. Symbolically:

$$a \times b = b \times a,$$

for any numbers a and b. Since it is customary to omit the times sign between letters, this is usually written $ab = ba$. (However, omitting the times sign between two numbers invites disaster. Just imagine the trouble you would cause by writing 3×5 as 35.)

When we memorize the multiplication table, the commutative law cuts our work in half. As soon as we know that 6×9 is 54, we also know that 9×6 is 54. So, instead of memorizing 81 products, we need memorize only 45, namely the squares of 1 through 9 and the 36 products where the first factor is less than the second. But the products $1 \times 1, 1 \times 2, \ldots, 1 \times 9$ come practically free of charge. That leaves, all told, only 36 products to memorize. Here they are:

	2	3	4	5	6	7	8	9
2	4	6	8	10	12	14	16	18
3		9	12	15	18	21	24	27
4			16	20	24	28	32	36
5				25	30	35	40	45
6					36	42	48	54
7						49	56	63
8							64	72
9								81

Memorizing these 36 products shouldn't loom as a giant challenge. It's comparable to remembering the lines of a small part in a play. Besides, the patterns in the products can serve as shortcuts. For instance, note that $6 \times 8 = 48$ is 1 less than 7×7, and that $5 \times 7 = 35$ is 1 less than $6 \times 6 = 36$. This pattern gives a way to remember the product of two numbers that differ by 2 if you have already memorized the squares: The product is 1 less than the square of the number that lies between them. (I admit that I have trouble with 6×9 and 7×8, having to remind myself which is 54 and which is 56.)

But the point of this chapter is not to heap praise on the multiplication table. Rather, it is to review the five things you can do with two numbers and the properties of these operations. Seeing them all in one place will show their essentials and their relations to each other. When they are introduced over a period of six years in school, it is hard to get this overall view.

Given two numbers, a and b, you can add, subtract, multiply, or divide them, obtaining, in order,

$$a + b, \ a - b, \ ab, \text{ and } a/b.$$

(If b is 0, the quotient a/b is not defined, as we see later in the chapter.) The quotient is also denoted

$$\frac{a}{b}.$$

The notation $a \div b$ survives mainly in the elementary grades and on calculators. I'll first discuss these four operations before going on to the fifth, which involves repeated multiplication.

Here are the basic properties of addition:

1. $a + b = b + a$.　(The commutative law)
2. $a + (b + c) = (a + b) + c$　(The associative law)
3. $0 + a = a$
4. For each number a there is a number, denoted $-a$ such that $a + (-a) = 0$.
   ($-a$ is called the *opposite* of a. For instance, the opposite of -3 is 3, and the opposite of 3 is -3.)

The associative law gets its name from the fact that b is *associated* with a on one side of the associative law and with c on the other side. This law permits us to discard parentheses and write simply $a + b + c$.

Subtraction is just the little brother of addition. When we ask, "What is $5 - 3$?" we are really asking, "What do we add to 3 to get 5?" In other words, we want to fill in the box in this addition:

$$3 + \square = 5.$$

Since $3 + 2 = 5$, we write $5 - 3 = 2$. When you pay $5 for a $3 item, the clerk may check the change by saying, "three," then counting out, "four, five," showing that the change, when added to the price of the item, gives the amount you put on the counter. That procedure translates the subtraction back into the original addition.

Unlike addition, subtraction has no attractive qualities. It is not commutative ($5 - 3$ is not equal to $3 - 5$). It is *not* associative: $7 - (4 - 2)$ does not equal $(7 - 4) - 2$, the first being 5, the second, 1. I find that I have to be very careful when working with subtraction, since it is such an awkward operation.

The third operation, multiplication, is as pleasant as addition. The rules it obeys are similar to those for addition. I already mentioned the first one.

1. $ab = ba$ (The commutative law)
2. $a(bc) = (ab)c$ (The associative law)
3. $1a = a$
4. For each number a, other than 0, there is a number, denoted $1/a$, such that

$$a \times \frac{1}{a} = 1.$$

($1/a$ is called the *reciprocal* of a.)

Note that 1 plays the role in multiplication that 0 plays in addition. Just as 0 has "no effect" in addition, 1 has "no effect" in multiplication.

Multiplication and addition are connected by the *distributive* law,

$$a(b + c) = ab + ac.$$

Though this law looks simple, it is quite complicated, for it involves three multiplications and two additions. If I rewrite it as

$$(b + c)a = ba + ca,$$

it makes sense. It's as though it says, "$b + c$ dogs is the same as b dogs plus c dogs." A more elegant way of looking at it is geometric, by considering the area of a rectangle cut into two smaller rectangles, as in Figure 6.

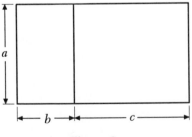

Figure 6

The area of the entire rectangle is its width times its length, $a(b + c)$. But this area is the sum of the areas of the two smaller rectangles, $ab + ac$.

The distributive law is the most important of the laws governing addition and multiplication. Disobeying this law causes thousands of errors every day in algebra classes across the nation.

The fourth operation, *division,* is the little brother of multiplication. When we ask, "What is 8 divided by 4?" we are really asking, "What must we multiply 4 by to get 8?" In other words, we are filling in the box in this equation:

$$4 \times \square = 8.$$

Since $4 \times 2 = 8$, we write $8/4 = 2$. Similarly, since $4 \times 1.75 = 7$, we also write $7/4 = 1.75$.

Observe that the symbol $7/4$ stands for the result of a division and also names a certain fraction. No trouble comes from this, because when $7/4$ denotes a fraction, it is true that

$$4 \times \frac{7}{4} = 7.$$

So the fraction $7/4$ does equal the result of dividing 7 by 4.

Like subtraction, division behaves in a lawless fashion. For instance, it certainly is not commutative, since $3/5$ does not equal $5/3$. The only attractive quality of division is that $a/a = 1$ when a is not 0.

Division by 0 is nonsense. This is why. To divide, say, 4 by 0, we must fill in the box in this multiplication,

$$0 \times \square = 4.$$

Since 0 times any number is 0, there is no way that we will ever fill in that box and get 4. So the symbol $4/0$ is meaningless. Oh, we can write it down, and the paper won't rebel. We can, in a similar fashion, write "the dog plays the violin," but that does not produce a musical dog.

The symbol $0/0$ is meaningless for another reason. To divide 0 by 0 we must fill in the box in the equation

$$0 \times \square = 0.$$

It's very easy to do that. For instance, $0 \times 3 = 0$, $0 \times (-7) = 0$, $0 \times (1/2) = 0$. The trouble is that it's too easy. Any number you please goes in the box. Therefore, the symbol $0/0$ does not name a specific number. If you ever run into it, you can be sure that something has gone wrong.

So much for the first four of the five things you can do with two numbers. To introduce the fifth, recall that multiplication, at least for whole numbers, can be viewed as repeated addition. For instance, 4×3 can be viewed as the sum of four 3's: $3 + 3 + 3 + 3$. The fifth operation, at least for whole numbers, can be viewed as repeated multiplication.

The product of four 3s, $3 \times 3 \times 3 \times 3$, is called "three raised to the fourth power." The standard notation for this product is

$$3^4$$

In this expression, 3 is called the *base* and 4 is called the *exponent*. As you may check, $3^4 = 81$. If b is any number and n is any whole number, the symbol b^n shall denote the product of n of the b's,

$$b^n = b \times b \times b \times \ldots \times b.$$

Here b is the base, n is the exponent; b^n is called an exponential. As you may check, $2^3 = 8$. One says, "Two to the three is eight," or "Two raised to the third power is eight," or "The third power of two is eight," or "The cube of two is eight." Note that b^2 stands for $b \times b$, the square of b, and b^3 stands for $b \times b \times b$, the cube of b. The symbol b^1 stands for b itself, since no multiplications are required.

The following table shows a few values of 2^n.

n	1	2	3	4	5	6	7	8	9	10
2^n	2	4	8	16	32	64	128	256	512	1,024

Note how rapidly 2^n grows. It doubles each time n increases by 1. This swift increase, called *exponential growth,* is a fair description of the dramatic growth of the human population in the last three centuries. That is one of the many applications of exponents.

Exponentiation, that is, the operation b^n, performed for any number b and any whole number n, has two convenient properties, which I will call the *sum law* and the *product law*.

The first is suggested by the fact that when you multiply 2 raised to the third power by 2 raised to the fourth power you get the product of seven 2s:

$$(2 \times 2 \times 2) \times (2 \times 2 \times 2 \times 2) = 2 \times 2 \times 2 \times 2 \times 2 \times 2 \times 2.$$

This shows why

$$2^3 \times 2^4 = 2^{3+4} \ (= 2^7).$$

(As a check, 2^3 is 8, 2^4 is 16, and 2^7 is 128, which is indeed 8 times 16.)

The *sum law* says that for any number b and any whole numbers m and n,

$$b^{m+n} = b^m \times b^n.$$

("b raised to a sum of two numbers is the product of b raised to the first number times b raised to the second number.") Observe that the sum law links exponentiation to a sum, $m + n$, and to a product. I will have much more to say about the sum law after I introduce the second law of exponents.

While the sum law is concerned with an exponent that is a sum, the product law is concerned with an exponent that is a product.

Consider the number

$$(2^3)^4.$$

(We don't need to know its value, but you may check that it is 4,096.) It stands for the "fourth power of the number 'two raised to the third power,'" that is,

$$(2^3)^4 = 2^3 \times 2^3 \times 2^3 \times 2^3.$$

But each factor 2^3 equals $2 \times 2 \times 2$. Thus,

$$(2^3)^4 = (2 \times 2 \times 2) \times (2 \times 2 \times 2) \times (2 \times 2 \times 2) \times (2 \times 2 \times 2). \qquad (1)$$

On the right side of equation (1) there are four groups of three 2s. That means there are 4×3 2s on the right side. Since 4×3 equals 3×4, there are 3×4 2s on the right side of (1). In other words,

$$(2^3)^4 = 2^{3 \times 4},$$

which is 2^{12}. The same reasoning gives us the general *product law* for any base b and any whole numbers m and n,

$$b^{mn} = (b^m)^n.$$

The product law connects three exponentials and one product.

So far we have defined 2^n only when n is a whole number. What should the symbol 2^0 mean? We are the masters, and we can give that symbol any meaning we want. However, it makes sense to define it in a way that is most convenient for us. Happily, what is simplest for us also turns out to be the most useful in the real world. As our guiding principle, let us try to have the sum law of exponents continue to hold, even when the exponents are not whole numbers. As an example, let's see what the sum law tells us that 2^0 should be.

It makes no sense to say "Take the product of zero 2s." But it does make sense to have the sum law hold when the exponent is the sum of 0 and 1. In that case we want

$$2^{0+1}$$

to equal

$$2^0 \times 2^1.$$

In short, we want

$$2^{0+1} = 2^0 \times 2^1$$

But $0 + 1$ equals 1, and 2^1 is 2. So the last equation becomes

$$2 = 2^0 \times 2.$$

That tells us that 2^0 times 2 must be 2. There is only one number that can do that, namely, 1. So, if the sum law of exponents is to hold when an exponent is 0, then 2^0 must be 1. We have no choice. For similar reasons we shall have b^0 equal to 1 for any base (other than the base 0). The symbol 0^0 is needed sometimes and is defined to be 1.

Let's see what happens when the exponent is negative. What should 2^{-1} be? The basic property of -1 is that when we add it to 1, we get 0: $(-1) + 1 = 0$. Wanting the sum law to hold when the exponent is $(-1) + 1$, we write

$$2^{(-1)+1} = 2^{-1} \times 2^1,$$

or, what amounts to the same,

$$2^0 = 2^{-1} \times 2^1.$$

Let's see if that will tell us what 2^{-1} should be. Using the fact that 2^0 is 1, and 2^1 is 2, we have

$$1 = 2^{-1} \times 2.$$

There is only one number whose product with 2 is 1, namely, one-half. For this reason we define 2^{-1} to be $\frac{1}{2} = 0.5$.

More generally, for any positive base b, we define b^{-1} to be the reciprocal of b. For instance, $4^{-1} = \frac{1}{4} = 0.25$ and $(\frac{1}{3})^{-1}$ is the reciprocal of $\frac{1}{3}$, which is 3.

What should b^{-2} be? The essence of -2 is that, when added to 2 it gives 0,

$$(-2) + 2 = 0.$$

If we want b^{m+n} to be the product $b^m \times b^n$ even when m is -2, then we must demand that

$$b^{(-2)+2} = b^{-2} \times b^2.$$

But $b^{(-2)+2} = b^0$, which is 1. So now we have

$$1 = b^{-2} \times b^2.$$

This forces b^{-2} to be the reciprocal of b^2. For instance,

$$5^{-2} = \frac{1}{5^2} = \frac{1}{25} = 0.04.$$

Similarly, whenever n is a whole number, we define b^{-n} to be the reciprocal of b^n

To review what we have accomplished so far, let's make a table of the values of 2^n for all the exponents n from -5 to 5. (I urge you to check my arithmetic.)

n	-5	-4	-3	-2	-1	0	1	2	3	4	5
2^n	0.03125	0.0625	0.125	0.25	0.5	1	2	4	8	16	32

Every time the exponent n goes up by 1, the value of 2^n doubles. Or, to put it another way, every time the exponent goes down by 1, the value of 2^n is cut in half. That pattern, all by itself, would have told us how we should define 2^0, 2^{-1}, 2^{-2}, and so on.

For a positive base b, what should $b^{1/2}$ be? Well, one thing we know about ½ is that ½ + ½ is 1. If we want the sum law to hold for fractional exponents, then we must have

$$b^{1/2} \times b^{1/2} = b^{1/2 + 1/2}$$

So

$$b^{1/2} \times b^{1/2} = b^1 = b.$$

That means that $b^{1/2}$ must be a square root of b. (We choose the positive square root to keep the value of the exponential positive.) For instance, $25^{1/2}$ must be 5. As another example, $2^{1/2}$ is the square root of 2, which is about 1.414.

I used the sum law of exponents to figure out what $b^{1/2}$ should be. You may wonder, "What if you use the product law instead? After all (½) × 2 = 1. That's just as fundamental as (½) + (½) = 1." Let's see what the product law tells us that $b^{1/2}$ should be. It sure would mess things up if it suggested something other than the square root of b.

To begin, we will want the product law to hold when the exponents are ½ and 2, that is,

$$(b^{1/2})^2 = b^{(1/2) \times 2}.$$

This equation says that

$$(b^{1/2})^2 = b^1.$$

Since $b^1 = b$, we see that $b^{1/2}$ should be a square root of b. Happily, that is exactly what the sum law for exponents told us.

What I find surprising is that squaring or cubing a number, or finding the reciprocal of a number, or finding its square root are all just special cases of exponentials. Even more astonishing is that trigonometry turns out to be the study of the sum and difference of two exponentials. This discovery of Euler depends on calculus and complex numbers.

It is possible to define the exponential operation when the exponent is any fraction. The sum and product laws will advise what the values should be. For instance, you might like to show why $8^{1/3}$ should be 2. Then what do you think $8^{2/3}$ should be? What about $16^{-1/2}$? If you have an exponential key on your calculator, you could compare your opinion with what the calculator displays.

That completes the survey of the five things you can do with two numbers. In a sense, everything grows out of addition, at least for whole numbers. First, multiplication is repeated addition. Then exponents begin with repeated multiplication. Subtraction and division are just different ways of looking at addition and multiplication. That, in a nutshell, surveys the essentials of arithmetic.

··· 18 ···

Some Sum

What happens when you start with 1 and add ⅔, then add $(2/3)^2$, then add $(2/3)^3$, and then add $(2/3)^4$, and so on, adding ever higher powers of ⅔, never stopping? Do these sums get very big? Do they ever go beyond 10? Beyond 100? After all, every time you add another number, the sum gets bigger. Or do these particular sums never get very big? This chapter shows that they stay fairly small. Better than that, it even finds the number that the sums get really near to. In Chapter 19, on banking, we put this discovery to practical use.

This chapter is written in a style to help you practice reading the language of mathematics. I play the roles of both author and reader. As a reader, I show what I do in order to understand what I am reading. My thoughts, as a reader, appear in *italics*.

I'm going to read mathematics, so I'll have paper, pencil, and calculator ready. I won't let this fellow slip anything past me.

The goal of this chapter is to show that if r is a number between 1 and −1, that is, $−1 < r < 1$, then

$$1 + r + r^2 + r^3 + r^4 + \ldots = \frac{1}{1-r}.$$

The three dots stand for "keep on adding more and more terms in the same way, such as r^5, r^6, and so on. The more you add, the closer the sum will get to $1/(1-r)$. The series $1, r, r^2, r^3, r^4, \ldots$ is called a *geometric series*, a name that goes back at least to the time of Plato. The number r is called its *ratio*.

What's he talking about? I'll try $r = \frac{1}{2}$, *for it's a number between* -1 *and* 1. *This bozo claims that*

$$1 + \frac{1}{2} + \left(\frac{1}{2}\right)^2 + \left(\frac{1}{2}\right)^3 + \left(\frac{1}{2}\right)^4 + \dots = \frac{1}{1-(1/2)}.$$

Let me see. I'll use my calculator:

$$1 + \frac{1}{2} = 1 + 0.5 = 1.5$$

$$1 + \frac{1}{2} + \left(\frac{1}{2}\right)^2 = 1.5 + \left(\frac{1}{2}\right)^2 = 1.5 + \frac{1}{4} = 1.75$$

$$1 + \frac{1}{2} + \left(\frac{1}{2}\right)^2 + \left(\frac{1}{2}\right)^3 = 1.75 + \frac{1}{8} = 1.875.$$

With just four terms I'm already up to 1.875. *He says that if I keep on adding I'll get closer and closer to* $1/(1-(1/2))$. *So what is* $1/(1-(1/2))$? *I'll work it out:*

$$\frac{1}{1-(1/2)} = \frac{1}{(1/2)} = 2.$$

Well, 1.875 *is fairly close to* 2, *so maybe he's right. But wait! If I keep on adding, won't my sums get bigger and bigger, so that they may even get bigger than* 2? *I'll add on a few more terms to see what's happening, say, four more terms. So I'll add on the sum*

$$\left(\frac{1}{2}\right)^4 + \left(\frac{1}{2}\right)^5 + \left(\frac{1}{2}\right)^6 + \left(\frac{1}{2}\right)^7,$$

which is

$$\frac{1}{16} + \frac{1}{32} + \frac{1}{64} + \frac{1}{128} = 0.03125 + 0.015625 + 0.0078125 + 0.00390625$$
$$= 0.05859375.$$

Adding this to my 1.875 *gives a grand total of* 1.93359375. *Well, that is pretty close to* 2. *There's no sign yet that the sums will go beyond* 2. *Anyway, there's a chance that the guy is right. I'll watch him like a hawk.*

But wait. He also said that r *could be negative. What happens then? I'll try* $r = -\frac{1}{2}$. *This time he is claiming that*

$$1 + \left(-\frac{1}{2}\right) + \left(-\frac{1}{2}\right)^2 + \left(-\frac{1}{2}\right)^3 + \dots = \frac{1}{1+(1/2)},$$

that is,

$$1 - \frac{1}{2} + \frac{1}{4} - \frac{1}{8} + \ldots = \frac{2}{3}.$$

I'll see if this is reasonable, using six terms:

$$1 - \frac{1}{2} + \frac{1}{4} - \frac{1}{8} + \frac{1}{16} - \frac{1}{32} = 1 - 0.5 + 0.25 - 0.125 + 0.0625 - 0.03125$$

$$= 0.65625.$$

Okay. That's pretty close to ⅔, which is about 0.666, so maybe he's right. I still have no idea why it may be true. He'll have to convince me.

First of all, take only the case when r is positive. The numbers r, r^2, r^3, r^4, ... get smaller and smaller, shrinking toward 0. For instance, 0.9 raised to the 50th power is about 0.005.

No surprise there. I saw this already in my calculations.

Now mark where the numbers 0, 1, r, r^2, r^3, r^4, ... lie on the number line, as in the following picture.

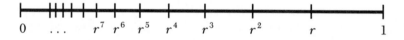

I don't know why he did this, but the picture looks right. This is my own picture for r = 0.8.

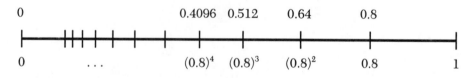

When r = 0.5, the following is my picture. The powers of 0.5 get really close to 0 much faster than do the powers of 0.8. Now I'm ready for his next message.

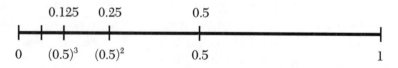

The segment from 0 to 1, which has length 1, is cut up into an end-less bunch of little segments. For instance, the first segment on the right, from r to 1, has length $1 - r$. The next segment, from r to r^2, has length $r - r^2$. The next has length $r^2 - r^3$ and so on. The following picture records these lengths.

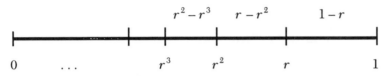

Since the length of the whole segment is 1, the sum of all the little lengths is 1. Writing these lengths in the same order that they appear in the picture, we have

$$\ldots + (r^2 - r^3) + (r^2 - r) + (1 - r) = 1.$$

Rewriting this equation backward, we have

$$(1 - r) + (r - r^2) + (r^2 - r^3) + \ldots = 1. \tag{1}$$

But $1 - r$ is a factor of each term. For instance, $r - r^2 = (1 - r)r$ and $r^2 - r^3 = (1 - r)r^2$. Factoring $1 - r$ out of each term of equation (1), we rewrite it as

$$(1 - r)(1 + r + r^2 + r^3 + \ldots) = 1. \tag{2}$$

Dividing both sides of equation (2) by $1 - r$ (which is not zero) gives us

$$1 + r + r^2 + r^3 + \ldots = \frac{1}{1 - r},$$

as we claimed at the beginning.

Neat trick. He drew a picture, and out it popped. But he missed the point. He didn't have to draw any picture at all in this example. It's much simpler. I stared at equation (1) for a while, and it became crystal clear. Practically everything cancels. The r and $-r$ cancel. The r^2 and $-r^2$ cancel, and so on,

$$(1 - \cancel{r}) + (\cancel{r} - \cancel{r^2}) + (\cancel{r^2} - \cancel{r^3}) + \ldots = 1.$$

All that is left is the first term, 1, which has nothing to cancel with. That's why all the terms on the left side of equation (1) add up to 1.

All that the fellow had to do was write down (1), factor out the $1 - r$, divide both sides by $1 - r$, and he would have been done. The picture isn't needed after all.

My argument is even better, for it works even when r is negative. His picture won't be of any use in that case. Poor guy, he missed the boat on this.

I wonder what would happen if we stopped the sum someplace. For instance, is there a short formula for

$$1 + r + r^2 + r^3 + r^4 ?$$

(Reader experiments a bit.)

Yes, there is. Using my method, I write down

$$(1 - r) + (r - r^2) + (r^2 - r^3) + (r^3 - r^4) + (r^4 - r^5).$$

Everything cancels except 1 and r^5.

$$(1 - \cancel{r}) + (\cancel{r} - \cancel{r^2}) + (\cancel{r^2} - \cancel{r^3}) + (\cancel{r^3} - \cancel{r^4}) + (\cancel{r^4} - r^5).$$

So I get

$$(1 - r) + (r - r^2) + (r^2 - r^3) + (r^3 - r^4) + (r^4 - r^5) = 1 - r^5.$$

Factoring out $1 - r$, just as before, and dividing both sides by it gives me

$$1 + r + r^2 + r^3 + r^4 = \frac{1 - r^5}{1 - r}.$$

The general pattern is clear: For any whole number k,

$$1 + r + r^2 + r^3 + \ldots + r^k = \frac{1 - r^{k+1}}{1 - r}.$$

I wonder why he didn't do this first? Well, it's just a matter of taste.

No, it is more than a matter of taste. I wanted readers who prefer to think geometrically to feel comfortable with the reasoning. Actually, with both your approach and mine available, any reader should be happy. For an even simpler geometric approach, you might turn to Chapter 32 now.

Now that we have the short formula for the sum of a geometric series, it's time to tell all the ways we use it. I already mentioned that it comes in handy in the next chapter, on banking. We need it in Chapter 28 to measure the steepness of a curve. In Chapter 30, it is the tool in finding the area under certain curves. Finally, in Chapter 31, it helps us show that the circumference of a circle is related to the reciprocals of all the odd whole numbers. Geometric series have many other applications, from determining the dosage of a medicine to the cost of a retirement pension.

Why didn't he tell me this at the beginning? I would have taken it more seriously. I like to know where I'm going. Still, better late than never. Hmmm . . . why should a circle have anything to do with all the odd whole numbers?

But I do have one thing more to say. I don't want the reader to think that whenever you keep adding up an endless supply of positive numbers that tend toward 0, their sums will always stay less than some fixed number.

The fourteenth-century French mathematician, Nicholas Oresme, wondered what would happen if he kept on adding up the reciprocals of the whole numbers, $\frac{1}{1}$, $\frac{1}{2}$, $\frac{1}{3}$, $\frac{1}{4}$, and so on, called the *harmonic series*. The first few sums are

$$\frac{1}{1} = 1.000$$

$$\frac{1}{1} + \frac{1}{2} = 1.500$$

$$\frac{1}{1} + \frac{1}{2} + \frac{1}{3} \approx 1.833.$$

When I first met this sequence I did some computations and bet someone that the sums stayed less than 13. I lost the bet. As Oresme showed, the sums get arbitrarily large. The following two diagrams show his reasoning.

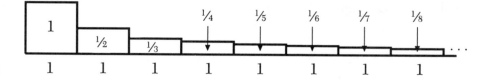

Each rectangle has width 1 and a height equal to the reciprocal of a whole number. The largest has area 1, the next has area $\frac{1}{2}$, the next has area $\frac{1}{3}$, and so on. We are wondering whether the total area of the endless staircase is finite or infinite. To show that it is infinite, inspect the next staircase, which fits inside the previous one.

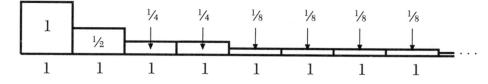

This staircase has 1 square of area 1, a rectangle of area $\frac{1}{2}$, 2 rectangles, each of area $\frac{1}{4}$, 4 rectangles, each of area $\frac{1}{8}$, 8 rectangles, each of area $\frac{1}{16}$, 16 rectangles, each of area $\frac{1}{32}$, and so on. Its total area is therefore

$$1 + \frac{1}{2} + \frac{2}{4} + \frac{4}{8} + \frac{8}{16} + \frac{16}{32} + \ldots,$$

which is

$$1 + \frac{1}{2} + \frac{1}{2} + \frac{1}{2} + \frac{1}{2} + \frac{1}{2} + \ldots,$$

an endless sum of ½'s. Since adding up more and more ½'s will take us beyond any number, the second staircase has an infinite area. Therefore, the first staircase has an infinite area. That means that the sum of the reciprocals of all the whole numbers is infinite. We should be grateful that the sum of a geometric series is finite.

But what about

$$\frac{1}{1 \times 2} + \frac{1}{2 \times 3} + \frac{1}{3 \times 4} + \frac{1}{4 \times 5} + \frac{1}{5 \times 6} + \ldots ?$$

I leave it to you to find what happens as you add more and more terms. Do the sums get to be gigantic or do they get nearer and nearer some number?

That there is no simple, routine way to decide whether the sum of a sequence of positive numbers is finite or infinite provides one of the many challenges that keeps mathematics alive and fascinating.

··· 19 ···

Out of Thin Air

I have often wondered where money comes from. After all, when our nation was an infant there was very little money. Now there are trillions of dollars. Finally I decided to find out and asked some economics professors. Their answer was so absurd that I thought they were pulling my leg. So I read an economics text, which, to my surprise, gave the same explanation as had the professors. Of course, the obvious way to create dollars is for the government to print currency and mint coins. But there is another way to manufacture money, and, to my delight, it is tangled up with a geometric series. It involves the banks.

It turns out that banks can make money appear out of nowhere. No magic. The trick is superior to counterfeiting for two reasons: It requires no printing press and it's perfectly legal. Instead, it depends on people's faith in the future. Here is how it works.

A bank has to maintain some reserves to meet the day-to-day demands of depositors. Think of these reserves as cash in the vault. If all the depositors feared that the bank was about to go broke and requested that their accounts be turned into cash, the bank might not have enough reserves to meet their demands. The bank would be embarrassed and the depositors angry. But if there is no such panic, the bank has enough reserves to satisfy the usual needs of its depositors.

Alas, sometimes all the depositors lose faith in the future and want to take their money out of the Bank of Faith. Then we have a run on the bank, or a panic. This happened in the Great Depression, and the federal government in March, 1933, declared a Bank Holiday, so the banks had a good excuse for not paying up: They were closed. In a few weeks, the solvent banks were allowed to reopen and faith gradually returned to the banking system.

Luckily, people usually have faith in their banks. They believe that if they wanted to turn their account into cash, they could. To help maintain this illusion, bankers dress conservatively and the architecture of a bank traditionally is reassuring, with marble floors, walls, and pillars, resembling a Greek temple. It would be more forthright to design a bank to resemble a Las Vegas casino, but that would destroy the illusion.

The Federal Reserve requires that a bank keep a certain fraction of a deposit as a reserve, but it lets the bank lend out the rest. For our purposes, let's say that a bank has to keep 20 percent, or ⅕, and can lend the remaining 80 percent, or ⅘. Watch carefully what happens, for this is the part I couldn't believe until I saw it in print.

Say that Joe Blow walks into the Bank of Faith and deposits $1,000. The bank keeps $200 and lends $800 (⅘ of 1,000) to Jane Doe. Jane deposits the $800 in the Bank of Faith or another bank. That bank keeps 20 percent of the $800 as a reserve but lends $640 (⅘ of 800) to Alice.

Already the magic show has begun. Joe thinks he has $1,000, Jane thinks she has $800, and Alice thinks she has $640. At the start, there was only Joe Blow's $1,000. Now, out of thin air, there is $1,000 + $800 + $640 = $2,440. That's a neat trick. Moreover, the banks earn interest on the money that they created. Clearly, banking is a very pleasant business.

Now, Alice deposits her $640. The bank then lends $512 (⅘ of 640) to Linus. Linus, in turn, deposits the $512, which permits the Bank of Faith to lend $409.60 (⅘ of 512) to Madeline. And so on. Let's see what the total amount of money is in the long run if this goes on and on: deposit, lend, deposit, lend, and so forth. Is it possible to create an infinite amount of money out of that initial $1,000? Let's see.

The deposit of $1,000 started the process. Then there was $800, which is (⅘)1,000. Next came $640, which is ⅘ of $800, that is,

$$\left(\frac{4}{5}\right)\left(\frac{4}{5}\right) 1,000 = \left(\frac{4}{5}\right)^2 1,000 = 640.$$

Then came $512, which is ⅘ of that amount, namely,

$$\left(\frac{4}{5}\right)\left(\frac{4}{5}\right)^2 1,000 = \left(\frac{4}{5}\right)^3 1,000 = 512.$$

At each stage the amount lent is ⅘ times the amount lent at the preceding stage. So the total amount lent by the banks is the endless sum,

$$1,000 + \left(\frac{4}{5}\right)1,000 + \left(\frac{4}{5}\right)^2 1,000 + \left(\frac{4}{5}\right)^3 1,000 + \left(\frac{4}{5}\right)^4 1,000 + \ldots.$$

The following picture shows the piles of money involved.

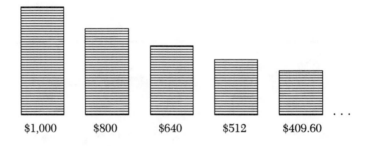

$1,000 $800 $640 $512 $409.60

We want to know whether this sum is finite or infinite. And, if it is finite, what is it? Factoring the 1,000 out of each term, we have

$$1,000 \left(1 + \frac{4}{5} + \left(\frac{4}{5}\right)^2 + \left(\frac{4}{5}\right)^3 + \left(\frac{4}{5}\right)^4 + \ldots \right).$$

The sum in parentheses is a geometric series, which we examined in Chapter 18. Its sum is

$$\frac{1}{1-(4/5)} = \frac{1}{1/5} = 5.$$

That means the total amount of money that Joe, Jane, Alice, Linus, Madeline, and so on think they have is $1,000 × 5 = $5,000. (Economists call 5 the *multiplier*. The multiplier depends on the percentage that must be kept on reserve. To be specific, it is the reciprocal of that fraction.)

Where there had been $1,000 there will be $5,000 in the long run. At least the Bank of Faith cannot make an infinite amount of money out of $1,000. That's reassuring. It pulled only $4,000 out of the hat.

One reader of this chapter thought all this was satire. "After all, whatever the bank lends has to be paid back, with interest." The loans do get paid back, even with more money created the way I described. I refer the unbeliever to any intermediate economics text.

There is another way to see how $1,000 turns into a total of $5,000. We won't need a geometric series. However, we will have to assume that the total amount is finite. That is a big assumption. We didn't have to assume that when we used the geometric series.

Call the total T. Since it is finite, we can do arithmetic and algebra with it. (We saw that it turns out to be $5,000, but pretend we don't know that.) T is made up of an initial deposit of $1,000 and all the deposits that followed. Call these later deposits *secondary*. Each secondary deposit is ⅘ a preceding deposit. The following picture describes the situation.

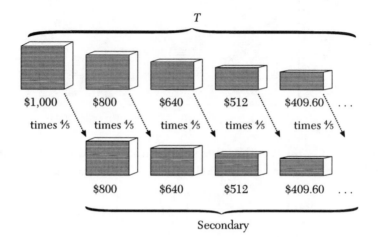

Secondary

The sum of all the secondary deposits is (⅘)T. The total deposit, T, is then the sum of the initial deposit of $1,000 and all the secondary deposits, (⅘)T. In other words,

$$T = 1,000 + \frac{4}{5}T.$$

All that's left is to solve this equation for T.

Not enjoying denominators, we multiply both sides by 5, obtaining

$$5T = 5,000 + 4T.$$

Subtracting $4T$ from both sides gives us

$$T = 5000.$$

Indirectly, then, we have added up all the deposits, showing that they total $5,000. That means we have found that

$$1 + \frac{4}{5} + \left(\frac{4}{5}\right)^2 + \left(\frac{4}{5}\right)^3 + \left(\frac{4}{5}\right)^4 + \ldots = 5.$$

In other words, the Bank of Faith has given us a new way to sum a geometric series. That's as magical as making money out of thin air. (The method works with any number r between 0 and 1, not just ⅘. Keep in mind, though, that in this approach we had to assume in advance that T is a finite number.)

One question may be on your mind, "Why can't I open a bank?" You will have to ask the economics professors. There must be a kit or handbook they could give you.

··· 20 ···

All There Is to Know about Fractions

When my younger daughter, Susanna, was in the fifth grade, it was clear that her teacher, though quite effective when teaching other subjects, did not enjoy mathematics. Often the mathematics session was shortchanged to make time for anything else. Sometimes it would disappear for days at a time. My offer to visit the class regularly to help out was gladly accepted.

To my astonishment, there were no physical objects in the room to illustrate the meaning of fractions—no measuring cups, no yardsticks, no so-called manipulatives at all. The learning was as abstract as in a graduate course in algebraic topology. Other than a few diagrams of the proverbial pie cut into a few equal pieces, there was little for the pupils even to look at. Ideally, there should be lots of props in the room, such as rulers and measuring cups. The least I could do as a start was make a *number line* and a *fraction kit*.

To make a number line, take a piece of paper about 10 feet long and 6 inches wide. Draw a long line on it, put a 0 near the center, a 1 about 2 feet to the right of the 0, and mark on the line where some of the famous integers and fractions are located. It might look like Figure 1.

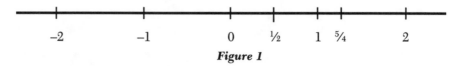

Figure 1

You could show fractions with denominators up through 5 or more. Even include a couple of fractions with a big denominator, such as $^{230}\!/_{95}$

and $^{231}\!/_{95}$, to show how close they can be. When this is pinned up on the wall, at home or at school, the children will see such things as

$$\frac{1}{2} = \frac{2}{4} = \frac{3}{6} \quad \text{and} \quad 1 = \frac{2}{2} = \frac{3}{3} = \frac{4}{4} = \frac{5}{5} = \frac{6}{6}.$$

These equations illustrate a concept that is at the heart of fractions: Different fractions can have the same value. I put negative numbers on the line to show that the line goes both ways, to the right and to the left.

Note that a fraction need not be less than 1, though in daily life it always is. In school, a fraction such as $\frac{7}{5}$, which is larger than 1, is called *improper*, as though the fraction had done something wrong. I have no impulse to change it to the *mixed fraction* $1 + \frac{2}{5}$, written $1\frac{2}{5}$. When you multiply fractions, it is usually more convenient to leave them as fractions, as you will see later. For instance, it is much easier to multiply $\frac{7}{5}$ by $\frac{8}{3}$ than $1\frac{2}{5}$ by $2\frac{2}{3}$.

A fraction kit is also easy to make, and I made one so that the pupils had something they could touch and move around. I cut a strip about 1 inch wide and 12 inches long out of tagboard. I put a 1 on it, to serve as the unit. Then I made two pieces, each 1 inch by 6 inches, out of a different color tagboard, and I labeled each of them ½. Similarly, I made three pieces 4 inches long, labeled ⅓, and four pieces 3 inches long, labeled ¼. I went through all the denominators up through 12. (Inconvenient denominators, such as 5 and 7, gave a little more trouble than the nice denominators, such as 6, 8, and 12, but I approximated them pretty well.) I assembled all of these pieces in a frame, as shown in Figure 2.

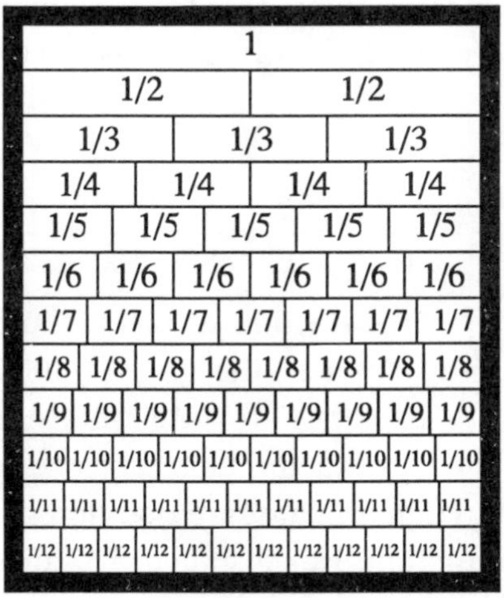

1											
1/2						1/2					
1/3				1/3				1/3			
1/4			1/4			1/4			1/4		
1/5		1/5		1/5		1/5		1/5			
1/6		1/6		1/6		1/6		1/6		1/6	
1/7	1/7	1/7	1/7	1/7	1/7	1/7					
1/8	1/8	1/8	1/8	1/8	1/8	1/8	1/8				
1/9	1/9	1/9	1/9	1/9	1/9	1/9	1/9	1/9			
1/10	1/10	1/10	1/10	1/10	1/10	1/10	1/10	1/10	1/10		
1/11	1/11	1/11	1/11	1/11	1/11	1/11	1/11	1/11	1/11	1/11	
1/12	1/12	1/12	1/12	1/12	1/12	1/12	1/12	1/12	1/12	1/12	1/12

Figure 2

Had I had more time, I would have had some of the pupils make the kit. Similar kits are available commercially, with fewer denominators.

With the kit on a desk, I asked a pupil questions such as:

Which is larger, ½ or ⅓?

Which is larger, ⅔ or ½?

Which is larger, ¾ or ⅚?

Which is larger, ⅝ or ⅔?

The pupil answered by looking at the board or by picking up the pieces and moving them about.

My next question was not so easy, "Which is larger, ⅝ or ⁷⁄₁₀?" The two are almost equal, as Figure 3 shows.

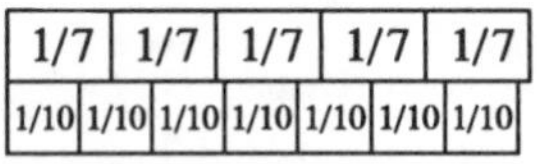

1/7	1/7	1/7	1/7	1/7

1/10	1/10	1/10	1/10	1/10	1/10	1/10

Figure 3

How could the pupil be sure of the answer? One way is to use a calculator, turning both fractions into decimals. But a pupil who is just learning fractions would not understand the display in a calculator, since it's a decimal. You have to know fractions *before* you study decimals. (At least you do in the United States, which does not use the metric system to measure lengths.) For instance, the decimal 0.35 is short for the fraction ³⁵⁄₁₀₀.

There is an easy way to decide, and it depends on the notion of *equivalent fractions,* or *equal fractions.* To introduce this notion, I asked, "Using the fraction kit, can you find other fractions equal to ½? Point them out on the board."

The pupils showed that

$$\frac{1}{2} = \frac{2}{4} = \frac{3}{6} = \frac{4}{8} = \frac{5}{10} = \frac{6}{12}.$$

Figure 4 shows how the fraction kit pieces illustrate this idea.

Equivalent fractions describe the same point on the number line. To put it in terms of arithmetic, they are different names for the same rational number. A *rational number* is a number that can be expressed as a fraction whose numerator and denominator are both integers. In particular, every whole number is a fraction: For instance, 13 can be written as the fraction ¹³⁄₁. (Chapter 21 faces the question, "Is every number a fraction?")

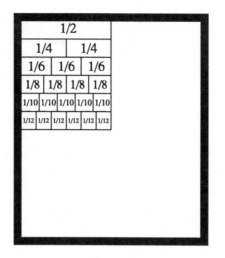

Figure 4

In daily life, we don't use the phrase *rational number*. We just use the word *fraction*. That is, we call a number that can be described by a fraction whose numerator and denominator are integers a *fraction*. This double use of the word fraction causes no difficulties, though it is a good idea to realize that we are using the word in two senses.

The pupils found several fractions equivalent to 1 or to ⅔. Then I felt they were ready to agree with this important principle: *Multiplying the numerator and denominator by the same number does not change the value of a fraction.* (Yes, it's perfectly fine to tell pupils something if they have been prepared to understand the message.) With the aid of this idea, you can then easily compare ⅝ and 7⁄10.

Write each of them so both have the same denominator. The simplest denominator to use is 7 times 10, or 70. We have

$$\frac{5}{7} = \frac{5 \times 10}{7 \times 10} = \frac{50}{70}$$

and

$$\frac{7}{10} = \frac{7 \times 7}{10 \times 7} = \frac{49}{70}.$$

Now it is easy to see that ⅝ is larger than 7⁄10, since 50 is larger than 49.

The fraction kit can also help in addition. A problem like finding

$$\frac{3}{7} + \frac{2}{7},$$

where the denominators are the same, is easy. Just add the numerators. ("Three dogs plus two dogs is five dogs.") The sum is $\frac{5}{7}$. But finding the sum

$$\frac{1}{2} + \frac{1}{3}$$

is not so quick. (There is the temptation to "add the numerators and add the denominators", getting $\frac{2}{5}$. But $\frac{2}{5}$ can't be correct: It's less than $\frac{1}{2}$, which is one of the two numbers being added.) The pupil may represent the sum by forming a "train" of two pieces, as in Figure 5.

1/2	1/3

Figure 5

The pupil moves this train along the fraction board until it matches an available length, as in Figure 6.

<table>
<tr><td colspan="3">1/2</td><td colspan="2">1/3</td></tr>
<tr><td>1/6</td><td>1/6</td><td>1/6</td><td>1/6</td><td>1/6</td></tr>
</table>

Figure 6

This shows that the sum is $\frac{5}{6}$. After a few more examples like this, the pupil may discover that the easiest way to add two fractions is to rewrite them so that they have the same denominators. (If the pupil doesn't notice the general principle, I would state it myself.) Without using the fraction kit, we have

$$\frac{1}{2} + \frac{1}{3} = \frac{3}{6} + \frac{2}{6} = \frac{5}{6}.$$

Clearly, the notion of *common denominator* is the key both to comparing the sizes of two fractions and to adding two or more fractions. I wouldn't worry about finding the *smallest* common denominator. There's no need for such elegance or economy. It's so easy to find a common denominator automatically: Just multiply the two denominators. Why make life unnecessarily tedious?

This is how I find

$$\frac{5}{6} + \frac{3}{4}.$$

I multiply 6 and 4, getting 24. Then I have

$$\frac{5}{6} = \frac{5 \times 4}{6 \times 4} = \frac{20}{24}$$

and

$$\frac{3}{4} = \frac{3 \times 6}{4 \times 6} = \frac{18}{24}.$$

From that, I find

$$\frac{5}{6} + \frac{3}{4} = \frac{20}{24} + \frac{18}{24} = \frac{38}{24}.$$

I admit that $^{38}/_{24}$ is not reduced, but so what? If need be, I could reduce it. Anyone offended by $^{38}/_{24}$ can reduce it to $^{19}/_{12}$ and be content. Finally, anyone allergic to improper fractions can write it as $1^{7}/_{12}$.

Of course, I could have used 12 as the common denominator in that addition, but that would have required more thought. And in doing arithmetic I want to be automated, not philosophical.

Subtraction is similar. For instance,

$$\frac{5}{6} - \frac{3}{4} = \frac{20}{24} - \frac{18}{24} = \frac{2}{24}.$$

What about multiplication and division of fractions? Oddly, both turn out to be simpler than addition and subtraction. In both cases, there is no need to change the denominators.

How to explain the multiplication of fractions? I would remind the pupils that 2×3 stands for "two of the threes." Then I might start with these questions:

What is ½ of ½?

What is ½ of ⅓?

What is ⅓ of ½?

These can all be answered with the help of the fraction kit. After a few more questions like these, one can state the general principle: The product of two fractions whose numerators are both 1 is the fraction whose numerator is 1 and whose denominator is the product of the two denominators.

But then I ask, "What is ½ of ⅗?" The pupil knows that

$$\frac{1}{2} \times \frac{1}{5} = \frac{1}{10}.$$

Then

$$\frac{1}{2} \times \frac{3}{5}$$

is three times as large. So

$$\frac{1}{2} \times \frac{3}{5} = \frac{3}{10}.$$

Then the pupil could find

$$\frac{7}{2} \times \frac{3}{5},$$

since it is seven times as large as ½ × ⅗. So

$$\frac{7}{2} \times \frac{3}{5} = 7 \times \frac{3}{10} = \frac{21}{10}.$$

After a few more examples like this, the pupil or I could state the general principle: *To multiply two fractions, multiply their numerators and multiply their denominators.*

As we see, multiplying two fractions is quite easy. Unfortunately, many children want to add two fractions by adding their numerators and adding their denominators. As we saw, addition of fractions is much more involved than that. To convince children that their hoped-for shortcut isn't right, ask them to use it to find the world's simplest fraction sum,

$$\frac{1}{2} + \frac{1}{2}.$$

They will get ²⁄₄, which equals ½. But a half plus a half is not a half.

After addition, subtraction, and multiplication comes division of fractions. The key to understanding division is a simple property of multiplication of fractions. I'll introduce this principle by an example.

What is

$$\frac{2}{9} \times \frac{9}{2}?$$

By the rule for multiplying fractions,

$$\frac{2}{9} \times \frac{9}{2} = \frac{18}{18} = 1.$$

Similarly,

$$\frac{1}{4} \times \frac{4}{1} = \frac{4}{4} = 1.$$

This just says that "a quarter of four is one," which should make sense on the number line. A few cases like these then lead to another basic principle: *The product of a fraction and that fraction "turned upside down" is just 1.*

Before discussing division of fractions, I'd like to take a moment to discuss division of integers. What do we mean by "6 divided by 2 is 3?" We are asking the question, "2 times what is 6?" In other words, we are asking what goes in the box in this equation:

$$2 \times \boxed{} = 6.$$

This is what we mean by "2 goes into 6 three times."

So what is $\frac{7}{10}$ divided by $\frac{2}{9}$? In terms of the number line or fraction kit, this is asking, "How many times can $\frac{2}{9}$ be laid off along $\frac{7}{10}$?" Figure 7 shows that it can be laid off a little more than three times.

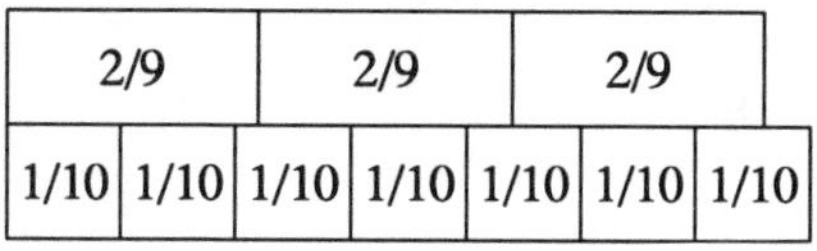

2/9	2/9	2/9

1/10	1/10	1/10	1/10	1/10	1/10	1/10

Figure 7

So $\frac{7}{10}$ divided by $\frac{2}{9}$ should be somewhere between 3 and 4. That will serve as a check on the computations we'll now be doing to get the exact answer.

We want to fill in the box in the equation

$$\frac{2}{9} \times \boxed{} = \frac{7}{10}.$$

To get the box all by itself, we must get rid of the $\frac{2}{9}$ that stands in front of it. To accomplish that, multiply both sides of the equation by $\frac{9}{2}$:

$$\frac{9}{2} \times \frac{2}{9} \times \boxed{} = \frac{7}{10} \times \frac{9}{2}.$$

But

$$\frac{9}{2} \times \frac{2}{9} = 1.$$

So we have

$$1 \times \boxed{} = \frac{7}{10} \times \frac{9}{2}.$$

Since 1 times any number is that number, we have

$$\boxed{} = \frac{7}{10} \times \frac{9}{2}.$$

This tells us that $\frac{7}{10}$ divided by $\frac{2}{9}$ is obtained by "turning $\frac{2}{9}$ upside down and multiplying $\frac{7}{10}$ by $\frac{9}{2}$." The answer is

$$\frac{7}{10} \times \frac{9}{2} = \frac{63}{20}.$$

Now $\frac{63}{20}$ is $3 + \frac{3}{20}$, which is a little over 3, in agreement with our observation at the start of the problem.

In short,

$$\frac{7/10}{2/9} = \frac{7}{10} \times \frac{9}{2} = \frac{63}{20}.$$

By the way, I would never, never write

$$\frac{7}{10} \div \frac{2}{9}.$$

The symbol $\div$ is obsolete in mathematics from algebra on up.

I find it surprising that division of fractions is easier than their addition. As long as we see why "turning a fraction upside down and multiplying" makes sense, then the mechanical process requires little thought.

If we know how to add, subtract, and multiply integers, we can easily add, subtract, multiply, and divide fractions. The key to addition and subtraction is putting the fractions over a common denominator. Only two principles have to be grasped. The first is that multiplying numerator and denominator by the same number does not change the value of a fraction. The second is that the product of a fraction and that fraction "turned upside down" is 1.

That's all there is to the arithmetic of fractions. There's no reason to make such a fuss over them.

··· 21 ···

Is Every Number a Fraction?

To place a satellite in orbit around the earth, you must give it a speed of about 17,717 miles per hour. However, to send a payload traveling to outer space and have it never fall back to earth, you must launch it with a speed of at least about 25,055 miles per hour. The ratio of these two numbers is 25,055/17,700, which is roughly 1.414. That number, 1.414, only approximates the ratio Escape speed/Orbit speed. Physicists, using calculus, have found that the actual ratio is $\sqrt{2}$, the square root of 2. Moreover, this same ratio holds for launches from any astronomical body, including the moon and the other planets.

Now, 1.414 is *not* the square root of 2, since the square of 1.414 is 1.999396. Close, but no cigar. In fact, since 1.414 ends in a 4, its square will end in a 6. That by itself tells us that the square of 1.414 could not be 2. For similar reasons, no decimal that stops can be the square root of 2. There is no chance that its square will have only 0s after the decimal point and 2 to the left.

Well, if $\sqrt{2}$ is not a decimal with a finite number of places, what kind of beast might it be? Could it be a fraction, that is, a number of the form m/n, where m and n are whole numbers? Such a number we call *rational* or, more customarily, a *fraction*. (More generally, any number that can be written in the form m/n, where m is an integer and n is a whole number, is called *rational*.)

We are asking whether there are whole numbers m and n such that $\sqrt{2} = m/n$. If there are such numbers, we could cancel any common factors.

So, if the square root of 2 is a fraction, there must be whole numbers, m and n, with no common factor larger than 1 such that $\sqrt{2} = m/n$.

Now, the basic fact about $\sqrt{2}$ is that its square is 2. That's really all we know about it. So we have

$$\left(\frac{m}{n}\right)^2 = 2.$$

It follows that

$$\frac{m^2}{n^2} = 2.$$

In order to simplify our lives, we get rid of denominators. In this case we multiply both sides of the last equation by n^2, obtaining

$$m^2 = 2n^2.$$

No longer do we have to think about square roots and quotients. We have moved to the simpler world of whole numbers and their products. Rather than asking, "Is the square root of 2 a fraction?" we are now asking, "Can twice the square of a whole number ever be a square of a whole number?" Though this doesn't sound at all like the question we asked at first, it amounts to the same thing.

Just to get a feel for the new question, let's run some experiments. Is twice 1^2 a square? No, it's 2. Is twice 2^2 a square? No, it's 8. This just misses being a square, since 9 is a square. A narrow miss, but in our question an inch is as good as a mile. You might try a few more cases. For instance, twice 5^2 is 50, which misses being a square, 49, by 1. However, as we saw in Chapter 15, even a million experiments will not settle the question, unless we happen to find a square whose double is also a square. Fortunately, the ancient Greek mathematicians showed beyond a shadow of a doubt that no one will ever find such a square. Their reasoning involves the notion of odd and even whole numbers. Let's see how they did it.

An odd whole number ends in 1, 3, 5, 7, or 9. So its square ends in 1, 5, or 9, as you may check. *Therefore, the square of an odd whole number is odd.* To put it another way, *if the square of a whole number is even, then that number must itself be even.* That's all we need. (If someone says to you, "I'm thinking of a whole number whose square is even," then you can say, "The number you're thinking of must be even.")

Now assume that m and n are whole numbers and that $m^2 = 2n^2$. As we already mentioned, we may take m and n to have no common factor larger than 1.

Because $2n^2$ has 2 as a factor, it is even. (Indeed, the definition of an even number is that it has 2 as a factor.) That tells us that m^2 is even. So m is even, as we just noted. That means that there is a whole number q such that $m = 2q$.

The equation $m^2 = 2n^2$ becomes

$$(2q)^2 = 2n^2,$$

which gives us

$$4q^2 = 2n^2.$$

Canceling 2 from both sides of this equation yields the equation

$$2q^2 = n^2.$$

This equation tells us that n^2 is even. Hence, n is even.

Now we have that both m and n are even. That means that they have 2 as a common factor. This contradicts our knowledge that their largest common factor is just 1. This contradiction must be due to some wrong assumption on our part. But the only assumption we made is that $\sqrt{2}$ is rational. We are forced to conclude that $\sqrt{2}$ cannot be written as a fraction whose numerator and denominator are whole numbers. In short, it's not rational.

That $\sqrt{2}$ is not rational does not complicate the launching of a rocket. Physicists and astronomers can approximate it by decimal expressions as closely as they wish. For instance, 1.414214 is good enough for all practical purposes. This is the fraction $^{1,414,214}/_{1,000,000}$ in disguise.

A number that is not rational is called *irrational.* Both the rational and the irrational numbers are called *real numbers.*

Once we have one irrational number, such as $\sqrt{2}$, we can manufacture an unlimited supply of irrationals. For instance,

$$\frac{7}{3}\sqrt{2}$$

is also irrational. For if it were rational, there would be whole numbers m and n such that

$$\frac{7}{3}\sqrt{2} = \frac{m}{n}.$$

Multiplying both sides of this equation by $^3/_7$ gives us

$$\sqrt{2} = \frac{3m}{7n}.$$

Since $3m$ and $7n$ are whole numbers, we would have the square root of 2 being rational. So our assumption that $(7/3)\sqrt{2}$ is rational must be wrong. Therefore, it is irrational.

The same argument shows that for any rational number r, other than 0, $r\sqrt{2}$ is irrational. Now, between any two rational numbers, no matter how close they are to each other, there are an infinite number of rational numbers. (Stop and convince yourself why this is so.) As we choose rational numbers r as close as we please to each other, the numbers $r\sqrt{2}$ are also as close as we please. Therefore, between any two numbers there are also an infinite number of irrationals. Both types intermingle everywhere on the number line, living in an intimate, complex tangle.

Even though the irrationals are so abundant, we never meet them in daily life. Prices are always rational; for instance, $9.37 is $937/100$. A carpenter, who may measure to the width of a saw blade, is content with a number like $2\,31/64$, which again is rational, being $159/64$. A batting average certainly is rational, since it is the number of hits divided by the number of at bats.

Up to about the year 500 B.C., mathematicians thought that all numbers were rational. The discovery by the Greeks that there are irrationals threw a monkey wrench into their reasoning. For example, when Archimedes (287–212 B.C.) developed his law of the lever, he had to break his argument into two cases, depending on whether the ratio of the two balancing weights was rational or irrational.

The Greeks thought of rational and irrational quite differently than we do. For them, arithmetic was geometry, and they viewed a number as the length of a line segment. Their basic concept was "one line segment measuring another line segment."

The segment AB *measures* another segment CD if we can lay off copies of AB along CD to match CD exactly. Figure 1 shows the case where AB measures CD, with 3 copies of AB matching CD.

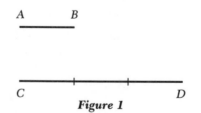

Figure 1

There are segments as short as we please that measure CD. For instance, a segment a millionth as long as CD also measures CD, since a million copies of it can be placed along CD.

Now think of two line segments, *CD* and *EF,* as in Figure 2.

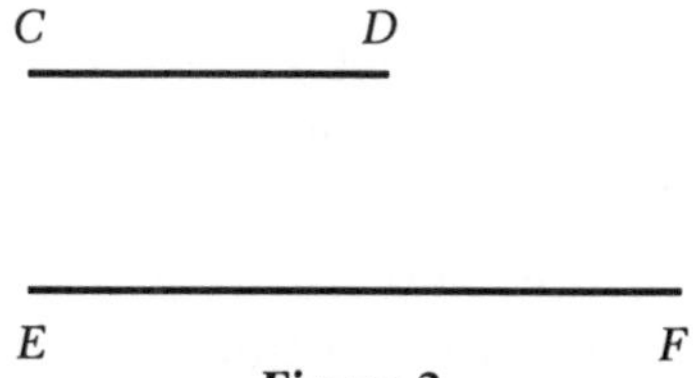

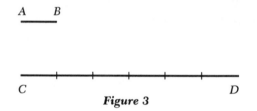

C   D

E   F

Figure 2

Is there necessarily a line segment *AB,* perhaps very short, that measures both segments *CD* and *EF*? Up to the year 500 B.C., mathematicians thought that there always would be such a common measure.

Let's translate these geometric ideas into the arithmetic of numbers.

When does a segment *AB* of length *r* measure a segment *CD* of length *a*? Answer: When there is a whole number *m* such that $mr = a$. Then *m* copies of *AB* fit along *CD,* as in Figure 3, where *m* is 6.

A   B

C   D

Figure 3

In this case, the length of *AB* is $a\!/\!6$. The lengths of segments that measure *CD* are $a\!/\!1, a\!/\!2, a\!/\!3, a\!/\!4, \ldots$, as small as you please.

Now say that there are two segments, one of length *a* and another of length *b*. For which numbers *a* and *b* will there be a common measure of both segments? If the length of the common measure is *r,* then there must be whole numbers *m* and *n* such that

$$a = mr$$

and

$$b = nr.$$

That tells us that

$$\frac{a}{b} = \frac{mr}{nr},$$

hence

$$\frac{a}{b} = \frac{m}{n}.$$

Even though a and b may be rather messy numbers, perhaps irrational, their quotient must be expressible in the form m/n, where m and n are whole numbers. In other words, *if segments of lengths a and b have a common measure, a/b must be rational.*

For instance, if $a = 10\sqrt{2}$ and $b = 23\sqrt{2}$, there is a common measure, namely, $\sqrt{2}$. And indeed, a/b is rational, being ¹⁰⁄₂₃.

Let's apply this to the hypotenuse and one leg of the right triangle in Figure 4.

By the Pythagorean theorem (developed in Chapter 22), $c^2 = 1^2 + 1^2 = 2$. Therefore, $c = \sqrt{2}$, and the ratio between the hypotenuse and one leg is $\sqrt{2}/1 = \sqrt{2}$, which is not a rational number. Therefore, there is no common measure for those two lengths.

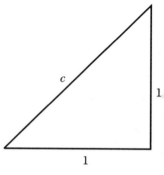

Figure 4

What if the ratio between the lengths of two segments is rational? Does that mean that they have a common measure? Let's see.

Call the lengths a and b. Assume that a/b is rational, that is, there are whole numbers m and n so that

$$\frac{a}{b} = \frac{m}{n}.$$

Multiplying both sides of this equation by bn to clear the denominators gives us

$$an = bm.$$

Then division by mn gives

$$\frac{a}{m} = \frac{b}{n}.$$

What does this equation tell us? It says, "If you cut a segment of length a into m equal pieces, and you cut a segment of length b into n equal pieces, then all these pieces have the same length." Call this common length r. A segment of length r measures both segments, so the segments of lengths a and b have a common measure.

The point of this detour is that the Greek notion of "having a common measure" is the same as our notion "the ratio of the lengths is rational."

Segments that lack a common measure appear often in geometry. Here is another example, which the geometer G. D. Chakerian called to my attention. First, draw a square, as in Figure 5.

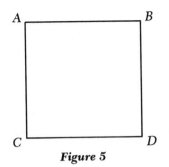

Figure 5

Now glue on a small rectangle to the right, but it cannot be just any rectangle. Starting with a very narrow rectangle along side BD, gradually increase the width until you obtain a rectangle $AEFC$ such that the little rectangle $BEFD$ that you add on has the same shape as the big rectangle $AEFC$, as in Figure 6.

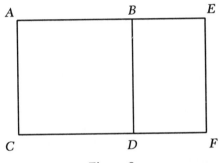

Figure 6

Rectangle $BEFD$ has the same proportions as rectangle $AEFC$, but it is rotated 90°. It is just a scaled-down version of the big rectangle $AEFC$.

Using only pictures, no arithmetic at all, let's show that the lengths AC and AE have no common measure.

To begin, assume that AC and AE do have a common measure r. Since $AC = AB$, r also measures AB. Figure 7 shows the situation.

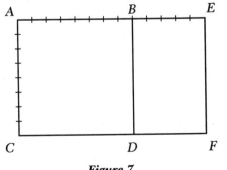

Figure 7

Therefore, r measures BE as well. It also measures EF, since AC and EF have the same length. So r measures both sides of the smaller rectangle $BEFD$. But $BEFD$ has the same shape as the rectangle $AEFC$ that we started with. So we can keep going, working next with rectangle $BEFD$.

Draw the square $BEGH$ in the rectangle $BEFD$. What's left over in $BEFD$ is a rectangle $HGFD$ similar to $BEFD$, as shown in Figure 8. This rectangle is a reduced image of the rectangle we started with, $AEFC$.

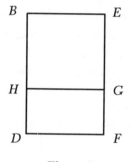

Figure 8

Since BE equals EG, r also measures EG, hence FG. Note that r also measures DF. So r measures both sides of the little rectangle $HGFD$. This reasoning could go on and on, producing ever smaller and smaller rectangles whose sides are measured by r. But this is absurd: Eventually the rectangles get so small that the lengths of their sides are smaller than r. So r couldn't possibly measure them. This shows that the sides of our original rectangle have no common measure. Hence, the ratio of those sides, whatever it is, is an irrational number. (Using algebra, you may show that the ratio of length to width is $(1 + \sqrt{5})/2$, our old friend from Chapter 6, the golden ratio, which has the habit of popping up in many branches of mathematics.)

In 1761, J. H. Lambert (1728–1777) proved that π is irrational. This implies that the diameter and circumference of a circle have no common measure. In other words, you can never find a circle whose diameter and circumference are each a whole number of inches long.

It is an amusing exercise to examine the following notion, which sounds like the very opposite of having a common measure. Call two lengths a and b *compatible* if there is some length c such that a measures c and b also measures c. If two lengths have a common measure, must they be compatible? If they are compatible, must they have a common measure?

In this chapter we explored the contrast between the rational and the irrational numbers. Chapter 26 looks into the question, "Are there just as many irrationals as there are rationals?" At first glance, the question sounds odd, since there are an infinite number of each type. But the surprising answer, though based on a notion so simple we usually meet it in kindergarten or first grade, has profound implications.

··· 22 ···

The Three Sides of a Right Triangle

A geoboard is a handy device, made of plywood and nails, for exploring the areas of figures whose borders are made up of straight segments. To make one, cut a square 14 inches on a side out of ½-inch plywood. Then draw dots on the board 1 inch apart, leaving a ½-inch margin along the edge. Finally, hammer a 2-inch finishing nail at each dot. The resulting geoboard should resemble Figure 1, which shows one in perspective and one from the top.

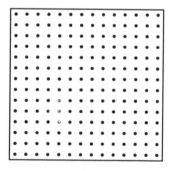

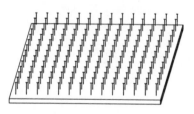

Figure 1

Hammering 196 nails may not be fun, but it does make a geoboard large enough to permit a wide variety of experiments. Since the border is half the distance between the nails, you could put several boards together, getting an even larger geoboard. By stretching rubber bands between the nails, you can then form polygons of all sorts of shapes and sizes.

As an alternative to a geoboard, you could draw regularly spaced dots on paper and then draw lines between them to form polygons. However, after each experiment you have to either erase the lines or start all over with a fresh piece of paper.

Now say that you form a polygon without dents on the geoboard, as in Figure 2.

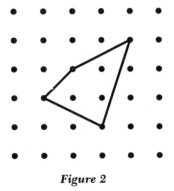

Figure 2

To find the area of such a polygon, first put a rectangular frame around it. Then break the area between the polygon and the frame into rectangles and right triangles with corners at nails, as in Figure 3.

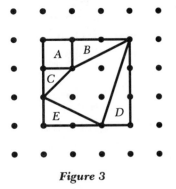

Figure 3

Since the area of a rectangle is just the product of its width and length, and the area of a right triangle is half the base times the height, it is easy to find the area outside the polygon. Then subtract this area from the area of the frame you made to find the area of the polygon. The calculations go like this:

Area of $A = 1 \times 1 = 1$.

Area of $B = (\frac{1}{2}) \times 1 \times 2 = 1$.

Area of $C = (\frac{1}{2}) \times 1 \times 1 = \frac{1}{2}$.

Area of $D = (\frac{1}{2}) \times 1 \times 3 = \frac{3}{2}$.

Area of $E = (\frac{1}{2}) \times 1 \times 2 = 1$.

Area of frame $= 3 \times 3 = 9$.

The area of the polygon is then

$$9 - \left(1 + 1 + \frac{1}{2} + \frac{3}{2} + 1\right) = 9 - 5 = 4 \text{ square inches.}$$

Admittedly, this is an indirect way to find the area. It is far more natural to stay inside the polygon and cut it up into rectangles and right triangles with all their corners at the nails. You're welcome to try. I couldn't do it.

After computing the areas of a variety of polygons, you will be ready to conduct the experiments that lead up to the famous theorem of Pythagoras that links the lengths of the three sides of a right triangle.

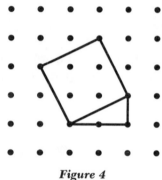

To begin, stretch one or more rubber bands to form a 1-by-2 right triangle. Then use some more bands to form the square based on the third side of the triangle. This, the longest side, is called the *hypotenuse,* from the Greek *hupo* (under) and *tenien* (stretch). The geometry is shown in Figure 4.

Figure 4

If you experiment, you will see that you will always be able to find two nails that complete the square. Then find the area of the square, with the aid of a frame, as shown in Figure 5.

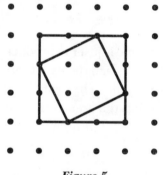

Figure 5

Since the frame has area 9, and we subtract four triangles, each of area 1, the square on the hypotenuse has area $9 - 4 = 5$ square inches. (Incidentally, you could also find this area by staying inside the square.)

This calculation can also be done for other right triangles formed on the geoboard. Using triangles small enough so that they and the square built on their hypotenuses fit on one geoboard, we might gather the following data:

First Leg of Triangle	1	1	1	1	2	2	2	3
Second Leg of Triangle	1	2	3	4	3	4	5	4
Area of Square on Hypotenuse	2	5	10	17	13	20	29	25

(The second column records the case we did.) Do you see the pattern that describes the area of the square in terms of the lengths of the two sides? If not, you could experiment with more triangles. (Don't look at the next paragraph. If this were a textbook, I wouldn't put it in.)

The pattern turns out to be, "The area of the square formed on the hypotenuse equals the sum of two numbers: the product of one leg times itself and the product of the other leg times itself." This is the Pythagorean theorem. I state it partly geometrically, in terms of the square, and partly numerically, in terms of products and sums, to present it in the form the experiments suggest.

To state the Pythagorean theorem completely in terms of geometry, we draw the squares formed on the two legs, as in Figure 6.

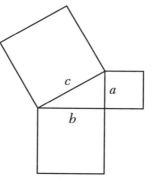

Figure 6

Now it reads, "The area of the square on the hypotenuse is the sum of the areas of the squares on the legs." In terms of the lengths a, b, and c shown in Figure 6, it reads,

$$c^2 = a^2 + b^2.$$

It is in this form that people usually remember the Pythagorean theorem.

Did the experiments done with the aid of the geoboard prove that the Pythagorean theorem holds for all right triangles? Not at all. However, there are simple arguments that show why it is always true. The simplest doesn't need any words whatsoever. It was discovered by Chinese mathematicians at least a thousand years ago and illustrates the old saying that one picture is worth a thousand words. In this case, the picture shows a square cut up in two ways, as in Figure 7. (The eight triangles are identical.)

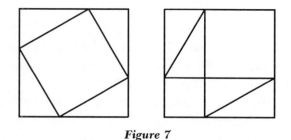

Figure 7

Gaze at it a while. Let it speak to you. Take your time, even several minutes. Without writing anything down, or doing any calculating, just by comparing areas, you will see why the area of the tilted square on the left equals the total area of the two smaller squares on the right. That proves the Pythagorean theorem.

As an illustration of this theorem, let's find the hypotenuse of the right triangle whose two legs are both 1. Calling the length of the hypotenuse *c*, we have

$$1^2 + 1^2 = c^2.$$

From this, it follows that $c^2 = 2$, and so *c* is the square root of 2, which is about 1.414.

The history of the Pythagorean theorem, like that of much of mathematics, is a blur. Babylonian mathematicians, a thousand years before Pythagoras, were already familiar with the theorem. They may have had a proof or may have just deduced it from experiments. For instance, one clay tablet shows a right triangle whose two legs are 30 and whose hypotenuse is

$$42 + \frac{25}{60} + \frac{35}{60^2}.$$

(They used base 60, not base 10.) If you stop to put this in decimal form and compare it with what the Pythagorean theorem gives for the hypotenuse, you would see that the two numbers are practically the same. Another tablet, which lists the three sides of a variety of right triangles, includes, for instance, the case when the two legs are 2,400 and 1,679 and the hypotenuse is 2,929.

Tradition has it that Pythagoras, who was active about 450 B.C., gave the first proof of the theorem. We do know that the theorem and its proof appear in Euclid's text, written about 300 B.C. In their book *Was Pythagoras Chinese?*, F. J. Swetz and T. I. Kao suggest, with some evidence, that the first proof may be the work of a Chinese mathematician a thousand years before Pythagoras.

Every day, thousands of engineers, scientists, mathematicians, carpenters, and calculus students apply the Pythagorean theorem. We call upon it in Chapter 31 when we obtain a way to find pi (π) using the reciprocals of all the odd whole numbers.

But as a more practical application, I now use it to answer the question: "When you're standing on top of Mount Shasta, how far can you see on the earth's surface?"

Taking the earth's radius as 4,000 miles, and the height of Mount Shasta as 3 miles, I make the sketch that shows the information (not to scale), as in Figure 8.

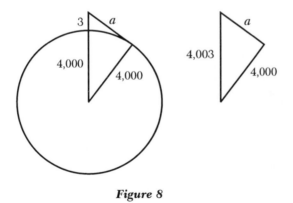

Figure 8

The distance sought, denoted a, is the length of one leg of a right triangle whose hypotenuse is 4,003 and whose other leg is 4,000. (The line of length a, being a tangent to the circle, is perpendicular to the radius shown.) So, by the Pythagorean theorem,

$$a^2 + 4,000^2 = 4,003^2.$$

Thus,

$$a^2 + 16,000,000 = 16,024,009.$$

From this it follows that

$$a^2 = 24{,}009,$$

and therefore,

$$a = \sqrt{24{,}009} \approx 155 \text{ miles.}$$

In short, you can see about 155 miles. Using the same approach, you can find the distance to the horizon when you're standing on a 100-foot-high cliff overlooking the ocean.

Let me end this discussion of the Pythagorean theorem with a couple of similar questions. First, if your eyes are 5 feet above the ground, how far can you see on the earth, which again we take as a smooth ball. Second, if, instead, you climb up on a ladder so your eyes are 10 feet above the ground, will you see about twice as far? As before, the Pythagorean theorem will help you answer the questions.

··· 23 ···

Pi Is a Piece of Cake—or Is It?

"**D**on't tell the pupils anything. Let them learn through discovery on their own." That's the prevailing pedagogical wisdom and it is appealing. But a teacher or parent who gives children a chance to carry out experiments and develop general principles themselves had better be prepared for surprises. I will illustrate this warning by describing my attempt to teach a bit of geometry through "discovery."

The distance around any circle is just a little more than three times the distance across the circle. In fancier terms, "The ratio between the circumference and the diameter is a little more than three." This ratio, denoted π (pi), is approximated pretty well by the decimal 3.14 and by the fraction $^{22}/_7$.

How should this notion be taught in school? Shirley Frye, when president of the National Council of Teachers of Mathematics in 1989, recommended that π should be introduced using jar lids. The pupils would measure the diameter and circumference of a lid, find the quotient, and record the data at the blackboard in three columns, labeled "Diameter," "Circumference," and "Ratio." She then asserted that "after the class collects the measurements, they will find that the number in column three is about 3.14. For all different size lids, the number will always be the same."

This approach sounds reasonable, even inevitable. In fact, I had used it years before, and I will explain what happened.

When one of my children was in the sixth grade, I was asked to visit the class and talk about mathematics. I decided to discuss the number π.

The choice seemed ideal: It gave the pupils a chance to carry out experiments and analyze data. Besides, it would easily fit into the time available.

Of course, I could skip the experiments and tell them flat out that the ratio is the same for all circles, and that, as a decimal, it's about 3.14. But if I used this approach, the pupils would quickly forget everything I told them.

The day before I was to appear, their teacher asked them to bring circular objects to class the next day. When I arrived I saw circles as small as a jar lid and as large as a bicycle wheel. I began by asking them to measure the distance across the circle, which they could do with a ruler. The distance around was not so easy to find. They could wrap a string around the circle and then measure the string or use a measuring tape.

Then, with me at the board acting as secretary, the class pooled their data, which went something like this:

Object	Diameter (inches)	Circumference (inches)	Ratio
Saucer	6.2	19.8	3.2
Garbage can lid	20.5	63.5	3.1
Wastebasket	9.3	29	3.1
Bicycle wheel	28	88.8	3.2
Jar lid	2.4	7.2	3.0

(It took a lot of work just turning the usual fractions of an inch into decimals. Too bad we don't use the metric system for measuring lengths.)

"What do you notice?" I asked.

"The bigger the circle, the bigger the circumference."

"Anything else?"

"The ratio goes from 3 to 3.2."

"What do you think is happening?"

No one offered a suggestion.

This is not what I had expected. I had hoped they would say that all the ratios would be equal were it not for experimental errors. They didn't, and what was I to do? My lesson was a shambles. Luckily for me, the bell rang and the class was over. If I were to return to give a follow-up lesson, I certainly would have them measure more accurately.

You may wonder, "What if the pupils, after measuring more accurately, still didn't suggest that the ratio of circumference to diameter is the same for all circles?" In that case, they would have had enough hands-on experience to be ready to absorb my message. Then I would tell them that were they able to measure perfectly, they would all get the same number no matter what size the circle, and that the number is an endless decimal that begins 3.14159.

When I came upon Frye's suggestion to use jar lids, I was surprised. Jar lids, being so small, give the least accurate estimates. Either she must have provided her pupils with special measuring devices, such as micrometers, or she never tried the lesson with a real class made up of real students.

In truth, though, it's hard to measure more accurately than the pupils had. Even working with large circles and precise measuring devices it would be difficult to get three or four decimal places of π. How, then, can we find π to ten or more decimal places? Clearly, there must be a way whose accuracy is not restrained by the limits on our ability to measure distances around jar lids, pots, pans, and bicycle wheels.

Even without any experiments we could say something about the ratio of the circumference to the diameter. The figure below shows a circle of diameter *D*, a square of side *D* around it, and a smaller, tilted square inside it.

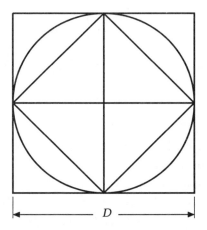

Since the perimeter of the big square is *4D*, the circumference of the circle is less than *4D*. This shows that π is less than 4. (Here we are using the fact that if a curve without bumps or dents lies inside another curve, it is shorter than the outside curve. If the inside curve is wiggly, this need not be true.)

Next we calculate the perimeter of the tilted square in order to find a number that is less than π. Call the length of a side of this square s, as shown below.

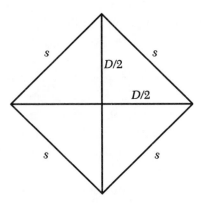

Now, s is the hypotenuse of a right triangle, both of whose legs have length $D/2$. By the Pythagorean theorem,

$$s^2 = \left(\frac{D}{2}\right)^2 + \left(\frac{D}{2}\right)^2.$$

Therefore,

$$s^2 = \frac{D^2}{4} + \frac{D^2}{4} = \frac{D^2}{2}.$$

It follows that s is the square root of $D^2/2$, which tells us that

$$s = \frac{D}{\sqrt{2}}.$$

So $4s$, the perimeter of the little square, is $4D/\sqrt{2} \approx 4D/1.414 \approx 2.828D$. Thus, π is larger than 2.828.

We now know that π is somewhere between 2.828 and 4. If you use hexagons instead, you will find that the inner hexagon has perimeter $3D$ and the outer one has perimeter $2\sqrt{3}D \approx 3.464D$. Therefore, π is between 3 and 3.464.

But how can we find π more precisely?

Archimedes, instead of using two squares, used two 96-sided regular polygons, one inside, one outside the circle. After a lengthy computation, he showed that π is between

$$\frac{25,344}{8,069} \text{ and } \frac{29,376}{9,347}.$$

(To four decimal places, these are 3.1409 and 3.1428.) Rather than report such awkward fractions, he said that π is between

$$3\frac{10}{71} \text{ and } 3\frac{1}{7}.$$

That is accurate enough to show that π is not a famous fraction.

It turns out that π is not a fraction, famous or not, as J. H. Lambert proved in 1761. You may be thinking, "Well, if it's not a fraction, not rational, at least I'd like to know what its decimal expression looks like. What is it?"

Consider what this question involves. The decimal system is based on the powers of ten. Why ten? Probably because we count using our ten fingers. When we write, say, 247, we are using a shorthand for

$$(2 \times 10^2) + (4 \times 10) + 7.$$

Similarly, 3.14 is shorthand for

$$3 + \frac{1}{10} + \frac{4}{10^2}.$$

So the question amounts to, "What is the relation between a circle and the fingers on my two hands?" When it is put that way, we would be surprised if there were any convenient relation, that is, a nice formula for the digits in the decimal form of π.

The decimal for π begins 3.14159265358979 . . . . To memorize it to six decimal places, use this mantra, "How I wish I could recollect pi," where the number of letters in each word equals a digit. If you need ten decimal places, remember, "How I wish I could recollect pi easily using one trick."

In 1989, Gregory and David Chudnovsky of Columbia University obtained over a billion decimal places of π. There are two reasons for such seemingly bizarre behavior; one practical, one theoretical.

First, the extensive calculation tests the capacity of a computer and uncovers "bugs" in its design. Second, the information obtained sheds light on an old question, "Are the digits of π random?" *Random* means that, in the long run, each of the ten digits tends to appear equally often, each about 10 percent of the time. Also, each of the hundred combinations of two digits tends to appear equally often, each about 1 percent of the time, and so on for triplets, quartets, and so forth. The first billion digits do suggest that the digits are random. But this information does not prove beyond a shadow of a doubt that the digits are really random. (Recall the moral of Chapter 15, which gives examples of wrong conclusions drawn from convincing data.)

How is it possible to compute π to so many decimal places? Certainly not by measuring larger and larger circles. Instead, mathematicians have developed formulas for computing it. One such formula is

$$\pi = 4\left(1 - \frac{1}{3} + \frac{1}{5} - \frac{1}{7} + \frac{1}{9} - \ldots\right),$$

which connects a circle to all the odd counting numbers. (Chapter 31 shows why this equation holds.) The more terms you use in the sum, the closer you can expect to get to π. If you use just three terms, the estimate is (to three places)

$$4\left(1 - \frac{1}{3} + \frac{1}{5}\right) \approx 3.467.$$

With four terms, you get

$$4\left(1 - \frac{1}{3} + \frac{1}{5} - \frac{1}{7}\right) \approx 2.895.$$

To get a better feel for these estimates, you might compute a few more. You will notice that they bob up and down, above and below π, as shown here.

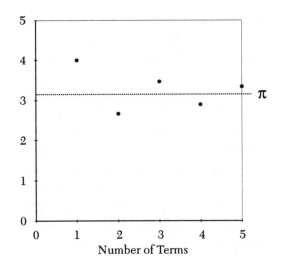

But even if you went out to 100 terms, your estimate would not yield three-decimal-place accuracy.

The Chudnovskys employed a much more complicated and efficient formula, related to the work of the Indian mathematician Srinivasa Ramanujan (1887–1920) at the turn of the century. Using a mere 100 of its terms, they could already obtain the first 1,400 digits of π.

Incidentally, one of the games in Chapter 15 can be used to estimate π, though not very efficiently. Recall the definition there of an S-number: a whole number whose factorization into primes involves no prime more than once. For any whole number n let n° (read as "n-star") be the number of S-numbers that are not greater than n. For instance, as you may check, when n is 100, n° is 61. Number theorists have proved that as you choose n larger and larger, n°/n, the fraction of the whole numbers up to n that are S-numbers, will get closer and closer to

$$\frac{6}{\pi^2}.$$

As an illustration, use $n = 100$. This gives the estimate

$$\frac{61}{100} \approx \frac{6}{\pi^2},$$

from which it follows that $\pi^2 \approx {}^{600}\!/_{61}$. Therefore, π is approximately the square root of 10, which is about 3.136.

It seems strange that a number defined in terms of a circle gets mixed up with the prime numbers. Even stranger, π appears throughout mathematics and its applications. To cite one more of its manifestations, statisticians see it in the formula for their *normal distribution,* the bell-shaped curve often used in assigning grades in large classes. In case you're curious, this curve is described by the expression

$$\frac{e^{-x^2/2}}{\sqrt{2\pi}}.$$

Here, e denotes every calculus student's favorite number, which is irrational, and whose decimal expression begins 2.718.

But let's return to the use of π in geometry. Once we know that the circumference of a circle is always π times its diameter and that π is about 3.14, we can find the area of a disk. For instance, let's find the area of a disk of diameter D, using an idea that goes back to the Greeks. Cut the disk into n equal pieces of pie. The figure on the next page shows the case $n = 20$.

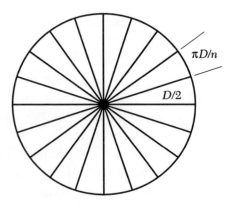

Each piece looks like a narrow triangle. Two of its sides are straight, while the very short one is curved. But when n is large, the pieces become very thin, and the curved side appears to be almost straight, as suggested by this figure:

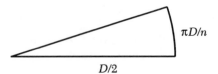

Each piece resembles a triangle of height $D/2$ and base $\pi D/n$. So the area of each piece is approximately

$$\frac{1}{2} \times \frac{\pi D}{n} \times \frac{D}{2} = \frac{\pi D^2}{4n}.$$

Since there are n of these pieces, their total area is about

$$n \times \frac{\pi D^2}{4n} = \frac{\pi D^2}{4}.$$

Therefore, as you choose n larger and larger, these estimates of the area of the disk get closer and closer to $\pi D^2/4$. So the area of a disk of diameter D must be

$$\frac{\pi D^2}{4} \approx \frac{3.14 D^2}{4} \approx 0.79 D^2.$$

That means that a disk occupies about 79 percent of the smallest square that holds it. (The irrigated disks in midwestern farms you see as you look down from a jetliner use about 79 percent of the land, as shown in the next figure.)

The first person to find the volume and surface area of a ball was Archimedes, who thought so highly of this work that he asked that a ball inscribed in a cylinder be displayed on his tombstone. He showed that the volume of a ball of diameter D is

$$\frac{\pi D^3}{6}.$$

Since $\pi/6$ is about 0.52, the ball occupies a bit over half of the volume of its enclosing cube, as shown in the figure below. (It occupies exactly two-thirds of the volume of a circumscribing cylinder.)

Archimedes also found the surface area of a ball. A look at the figure shows that the area of the top half of the surface is greater than the area of its circular shadow. So is the area of the bottom half. Therefore, the entire surface area is greater than twice that of the shadow. Archimedes showed that it is exactly four times the area of the shadow. This relation provides an easy way to recall the formula for the surface area: It is 4 times $\pi D^2/4$, which is simply πD^2.

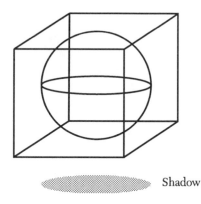

Shadow

Some 2,000 years after Archimedes, A. L. Cauchy (1789–1857) generalized this discovery. He showed that the surface area of any solid that has no dents or bumps—such as an egg or a box—is exactly four times the average of its shadows made in all possible directions. This fact has been used to build an automatic electronic lemon sorter.

Scientists, in an attempt to contact intelligent life in outer space, have transmitted the digits of the decimal form of π as a signal that life, intelligent or not, thrives on at least one planet. Surely any advanced civilization would meet this important number, even if its environment is so liquid that the inhabitants never see a circle and never wonder about the ratio between its circumference and its diameter.

That π has many uses should not come as a surprise. After all, the number 2 also has many uses: It can tell us how many wheels there are on a bicycle, the length of a Congressional term, or the number of hydrogen atoms in a water molecule. Had we studied statistics before geometry, we would be surprised that a number that appears in the normal distribution is related to the distance around a circle. That the number π appears in so many branches of the tree of mathematics simply reflects the underlying unity of the discipline.

When we think of π, let's not always think of circles. It is related to all the odd whole numbers. It also is connected to all the whole numbers that are not divisible by the square of a prime. And it is part of an important formula in statistics. These are just a few of the many places where it appears, as if by magic. It is through such astonishing connections that mathematics reveals its unique and beguiling charm.

After reading this chapter and discussing it with me, my wife, the poet Hannah Stein, wrote the poem on the next page, which describes not only her amazement at the many roles π plays, but my own as well.

••• Loving a Mathematician •••

The ether, or whatever's up there—
some infinite glassy staircase—
crackles for you
with truth, with beauty—and I
have never followed you even to
the second rung. I used to think Pi
was just a way of measuring circles.
You tell me now that Pi lurks
in gaseous, in liquid universes
where there are no circles, where rings
couldn't form if I dropped a pebble.
For there are no pebbles either—
no discs no balls no equators—
only pure structure.
It's true, you say,
that Pi always turns up,
like an old irrational uncle
who's been traveling around the country
doing card tricks. But circles
are only one of his arts:
Pi rolls his thumb through the ink
of odd numbers; from his hiding place in
square roots under square roots like
a wagon load of deviant potatoes
Pi intones: *dividing by*
a squared prime
has nothing to do with roundness.
Pi shines traces beyond
the galaxies mathematicians map,
Pi haunts the void between electrons
stalks black holes and red shifts.
Inching like a growing crystal into
the cosmic chinks, Pi waits
for thought to close in, waits
to be pounced on with a pencil
as his secrets repercuss
into patient, searching minds.
I ask you this: does Pi buckle
the whole universe together?
Can Pi be God?

For the first time I believe
I could follow you up and up—

··· 24 ···

Turning an Equation into a Picture

The first half of the seventeenth century witnessed a truly important wedding, far more important than the union of royal dynasties. For it was then that algebra and geometry became one. Two French mathematicians, René Descartes (1596–1650) and Pierre Fermat (1601–1665), were responsible for this enduring marriage. Descartes showed how to translate geometry into algebra, and Fermat showed how to translate algebra into geometry. Today, Descartes gets most of the credit because he wrote a book on his idea. However, in this chapter I focus on Fermat's point of view, since we use it in later chapters.

His idea is quite simple, as are many big ideas—such as continental drift or the helical form of a chromosome—once someone manages to think of them for the first time.

To start, draw two perpendicular lines on a piece of paper. Think of these lines as endless, even though the paper must stop. One line is horizontal, that is, parallel to the bottom of the page. The other is vertical, parallel to a side. For over three centuries the horizontal line has been called the x axis and the vertical line, the y axis. Figure 1 continues this tradition.

Turn the axes into identical number lines, as in Figure 2.

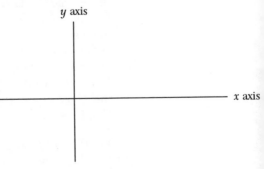

Figure 1

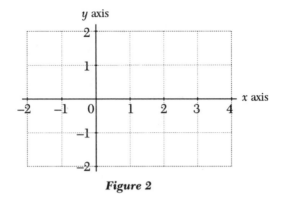

Figure 2

Every point on each axis corresponds to a number. Any point *P* in the plane can be described by a pair of numbers. To obtain those two numbers, draw a line through *P* parallel to the *y* axis and a line through *P* parallel to the *x* axis, as in Figure 3.

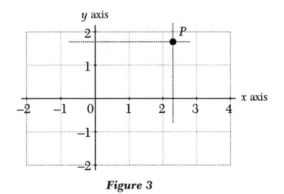

Figure 3

The vertical line through *P* meets the *x* axis at a point described by some number *x*. (In Figure 3, *x* is about 2.3.) This number is called the *x coordinate* of *P*. It tells how far the point *P* is from the *y* axis. For a point *P* to the right of the *y* axis, *x* is positive; if *P* is to the left of the *y* axis, *x* is negative.

The horizontal line through *P* meets the *y* axis at a point described by some number *y*. (In Figure 3, *y* is about 1.7.) This number is called the *y coordinate* of *P*. It tells how far *P* is from the *x* axis. For points above the *x* axis, *y* is positive; for points below, it is negative.

It is customary, then, to refer to *P* simply as the "point (*x*, *y*)," where *x* and *y* are called its *coordinates*. For instance, the point *P* in Figure 3 might be (2.3, 1.7). Figure 4 shows a few more points scattered about the so-called *xy* plane.

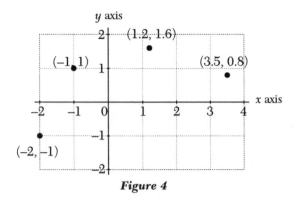

Figure 4

Now we come to Fermat's way of turning an equation that involves the letters x and y into a picture. I will illustrate it by the equation

$$y = x^2,$$

which will appear in later chapters.

The coordinates of most points do not satisfy the equation. For instance, the point $(2, 11)$ does not. When you replace x by 2 and y by 11 in the equation $y = x^2$, you get $11 = 2^2$, or $11 = 4$, which is not true. However, the point $(5, 25)$ does satisfy the equation because 25 does equal 5^2. The picture, or *graph*, of the equation is made up of all the points whose coordinates satisfy the equation. That's how Fermat turns the equation into a picture.

To find points on the graph of $y = x^2$, pick a number to be x. Then y is the square of that number. For instance, if you pick x to be 3, then y is 9. So the point $(3, 9)$ lies on the graph. So does $(-1, 1)$. Figure 5 shows seven points found in this manner.

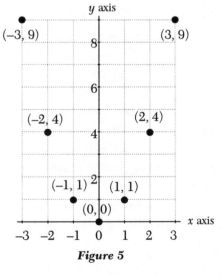

Figure 5

To save space, I kept x between -3 and 3. If I had let x be 4, then y would be 16, and the graph would use up half a page. Also, for convenience, I used as choices of x only integers. But all the numbers that lie between the integers are available as well. For instance, if $x = \frac{1}{2}$, then $y = \frac{1}{4}$; so the point $(\frac{1}{2}, \frac{1}{4})$ is also on the graph of the equation $y = x^2$. So is the point $(\sqrt{2}, (\sqrt{2})^2) = (\sqrt{2}, 2)$.

The collection of all the points (x, y) for which y is the square of x forms a smooth curve, as shown in Figure 6.

This curve, called a *parabola,* has been studied for over 2,000 years. Greek mathematicians discovered a remarkable property of this curve. To describe this property, imagine that the curve is made out of a reflecting material, like a mirror or shiny metal. It turns out that all the rays of light parallel to the y axis, coming from above the parabola, bounce off the curve and pass through a single point on the y axis, $(0, \frac{1}{4})$, denoted F in Figure 7. (This is easily established with the aid of calculus.)

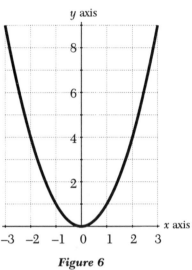

Figure 6

If you spin the parabola around the y axis to form a shiny bowl, you obtain a solar cooker. Since all the incoming light passes through point F, that is the place to put the hamburger or tofu.

On the other hand, imagine placing a lightbulb at point F. Then all the light from the bulb, after bouncing off the parabola, ends up parallel to the y axis. That is why reflectors in flashlights or headlamps are shaped like parabolas. Galileo (1564–1642) showed that the path of a thrown ball is parabolic. The cable holding up a uniform horizontal load, such as in a bridge, is a parabola. (However, a uniform rope or cable hanging from its two ends, a clothesline, for instance, does not

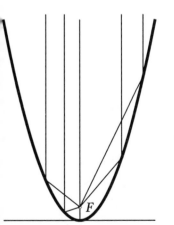

Figure 7

take on a parabolic curve. Instead, like the great arch in St. Louis, it has the shape of a catenary. The graph of $y = 2^x + 2^{-x}$ is an example of a catenary. Note that here the base is fixed and the exponents vary. That is quite a contrast to the equation $y = x^2$, where the exponent is fixed and the base varies.)

What does the graph of the equation $y = 2x$ look like? To find out, choose some values of x, find the related values of y, and then graph the points (x, y) found that way. This table puts the arithmetic in a neat form:

x	0	1	2	3	−1	−2
$y = 2x$	0	2	4	6	−2	−4

The table gives us six points, which are graphed in Figure 8.

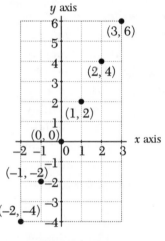

Figure 8

The six points lie on a straight line. If you graph more points that satisfy the equation $y = 2x$, you will find that they all lie on that line. In short, the graph of the equation $y = 2x$ is a straight line.

If you have a calculator with an exponential key, you might like to graph the equation $y = 2^x$. This graph does not resemble either of the graphs I have discussed. It is an example of graphs met in the study of populations that increase at a rate proportional to the size of the population. You might also enjoy sketching the graphs of $y = x^3$ and $y = \frac{1}{x}$, which look quite different from those already mentioned.

In all of the equations discussed so far, one side of the equation consists just of y, all by itself. But there is no need to make such a restriction. For instance, you might like to sketch the graph of the equation

$$x^2 + y^2 = 25.$$

It turns out to be a very famous curve that you see every day. Be sure to allow x and y to be positive, negative, or zero.

Why Negative Times Negative Is Positive

"Why is –1 times –1 equal to 1 and not –1?" This question is often answered with, "It just is, and stop asking such questions." This reply suggests that a mystery is involved, too deep for most humans to comprehend. There is no mystery and there is nothing deep at all. Here, in this chapter, I give three separate explanations from three different perspectives. All three advise us that –1 times –1 should be 1.

These explanations are based on two principles: "Let's keep life simple"; and "We, not the numbers, are the boss." Though neither is profound, both are useful, as we will soon see. When we defined exponentials, in Chapter 17, we already applied both principles.

My first explanation involves the *distributive rule*. Though we discussed this rule in Chapter 17, I'd like to review it, for it is the great bridge that links addition and multiplication. Figure 1, which shows a rectangle cut into two smaller rectangles, presents this rule pictorially.

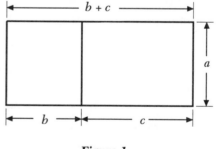

Figure 1

Since the area of a rectangle is the product of its width and length, the areas of the two smaller rectangles are the products ab and ac. The area of the whole rectangle is $a(b + c)$. Because the area of the whole rectangle is the sum of the areas of the smaller rectangles, we have

$$a(b + c) = ab + ac.$$

This equation, which involves two additions and three multiplications, is the *distributive rule*.

Figure 1 shows that the distributive rule holds when the numbers a, b, and c are all positive. To keep life simple, we want the same rule to hold when one or more of those three numbers are negative. We are the bosses, and we don't want to memorize more rules than necessary. With these lofty principles to guide us, let's see what −1 times −1 should be.

First of all, let's agree that 0 times any number is 0, and 1 times any number is that number. For instance, $(1) \times (-1) = -1$. A positive number times a negative number should logically be negative. (In football, a loss of 2 yards is recorded by −2. If this happens on each of three downs, the total effect is recorded by −6. So $3 \times (-2) = -6$. The same logic applies to a checking account.)

Now, the key property of −1 is that if you add it to 1 you get 0. So it's reasonable, perhaps inevitable, that we start with the equation

$$(-1) + 1 = 0.$$

In order to bring the distributive rule into action, I'll multiply both sides of that equation by −1, getting

$$(-1)[(-1) + 1] = (-1)0.$$

Since we want the distributive rule to hold for all numbers, whether they are positive or negative, we must have

$$(-1)(-1) + (-1)1 = (-1)0.$$

Now, $(-1)1 = -1$, since 1 times any number is that number, and $(-1)0$ is 0, since 0 times any number is 0. So we have

$$(-1)(-1) + (-1) = 0.$$

Adding 1 to both sides of this equation cancels the −1 on the left side and, on the right side, replaces 0 by 1. That gives us the equation

$$(-1)(-1) = 1.$$

In short, if the distributive rule is to hold for all numbers, then −1 times −1 must be 1.

As you may check, the same reasoning shows that the product of any two negative numbers must be positive. (Take the case $(-2)(-3)$ as a typical example and multiply both sides of the equation $2 + (-2) = 0$ by −3.)

The distributive rule, together with our goal of simplicity and the knowledge that we are the bosses, guided us to make the product of two negative numbers positive. But could some other rule suggest to us that it should be something else? If so, that would be a mess. Happily, whatever approach we choose for defining the product $(-1)(-1)$, it always guides us to make it 1, not -1.

For instance, let's see what the rules for exponentials tell us. For whole numbers x and y, we have the rule

$$(2^x)^y = 2^{xy}.$$

(This rule was discussed in Chapter 17.) Let's say that this rule is to hold even when both exponents, x and y, are -1.

Then we must have

$$(2^{-1})^{-1} = 2^{(-1)(-1)}. \tag{1}$$

But 2^{-1} is, as we saw in Chapter 17, the reciprocal of 2, that is, $\frac{1}{2}$. So $(2^{-1})^{-1}$ is the reciprocal of $\frac{1}{2}$, which is 2. That is,

$$(2^{-1})^{-1} = \left(\frac{1}{2}\right)^{-1} = 2.$$

Now, 2 is equal to 2^1. So equation (1) reduces to the equation

$$2^1 = 2^{(-1)(-1)}.$$

This last equation suggests, once again, that -1 times -1 should be 1.

My third explanation of why -1 times -1 should be 1 involves a graph of an equation. The graph of the equation $y = 2x$ is a straight line, as we saw in Chapter 24. If you experiment with a few positive values of the number a, you will see that the graph of $y = ax$ is a straight line. So, to keep life simple, we would like the graph of $y = ax$ to be a straight line even when a is negative. To be specific, let's look at the graph of the equation $y = (-2)x$.

To begin the sketch, I choose a few positive values of x and calculate corresponding values of y, which is (-2) times x. For instance, when x is 2, y is $(-2) \times 2 = -4$.

x	1	2	3	4
$y = (-2)x$	-2	-4	-6	-8

These four values of x and of y give four points on the graph, as shown in Figure 2. The graph of $y = (-2)x$, for positive x, is shown in Figure 2.

The graph of $y = (-2)x$, for positive x, is part of a straight line. If we want it to continue along that line when x is negative, what must y be when x is -3, for instance? As you travel up the line, you go up 2 units for each unit that you go the left. So if you move 3 units to the left of the 0, you should rise another 6 units. In other words, the point $(-3, 6)$ should also be

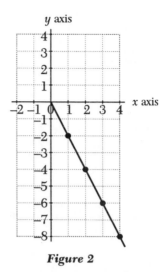

Figure 2

on the graph. That means that when x is -3, then the product $(-2)(-3)$ should be 6, a positive number. Then the graph of $y = (-2)x$ for all x, positive or negative, will be a single straight line, as shown in Figure 3.

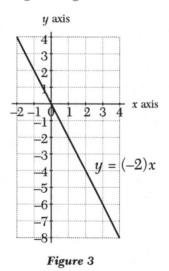

Figure 3

That will help keep our lives simple. Once again, "negative times negative should be positive." In a crude shorthand, $- \times - = +$.

All three approaches led us to the same conclusion, whether guided by the distributive rule, properties of exponentials, or the graph of an equation. This all took place in the self-contained world of mathematics. But, as I will now illustrate with the aid of a teeter-totter, even when "negative times negative" is met in studying the physical world, it should be positive.

Figure 4 shows a teeter-totter that balances at point F. Part of a number line, with its 0 at F, is drawn on the board to show distances from F.

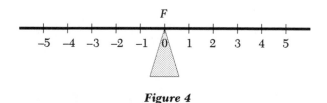

Figure 4

A position on the board can be positive or negative. Furthermore, a person can push up or down on the board. Physicists call an upward force positive and a downward force negative. The force can have a counter-clockwise or clockwise effect, depending on where it is applied. The farther the force is from F, the more turning tendency it produces. Physicists use the product of the force and the coordinate of the place where it is applied to measure this turning tendency:

Turning tendency = Force times position.

For instance, an upward force of 6 pounds at the point with coordinate 4 causes a turning tendency of 6×4. A downward force of 6 pounds applied at -4 produces a turning tendency $(-6) \times (-4)$. But a glance at Figure 5 shows that these two tendencies should be the same, both being counterclockwise.

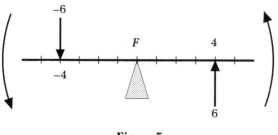

Figure 5

Thus, the physicists are happy that $(-6) \times (-4)$ is equal to 6×4, which is positive.

In the world of mathematics or in the world around us, negative times negative should always be positive. In theory or in practice, that choice is always the best. I would be quite surprised if anyone ever found a situation suggesting that the product should be negative. Whoever wants the product to be negative would have to invent a brand new number system.

··· 26 ···

A Fresh Look at Kindergarten

It was in kindergarten or first grade that most of us began our mathematical journey. It was there that we learned the meaning of the symbols 1, 2, 3, 4, 5, 6, 7, 8, and 9. For instance, our introduction to the idea of 2 meant looking at pictures like Figure 1, which shows two apples.

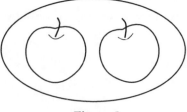

Figure 1

Another picture may have shown two bananas, as in Figure 2.

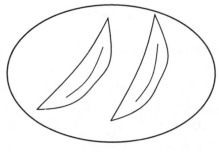

Figure 2

Then the teacher asked us to show that there are just as many bananas as apples. To do this, we matched each apple with a banana by drawing a line between them, as in Figure 3.

When we could pair off all the apples with all the bananas in this way, we said that there were just as many bananas as apples. To put it more formally, let us now say "the set of apples is co-numerous with the set

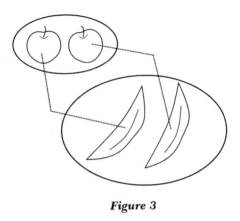

Figure 3

of bananas." This brings us to the key concept of the chapter.

Let us agree to call two sets *co-numerous* if we can pair off all the objects in one set with all the objects in the other set. That is a very precise definition, and we will abide by it scrupulously. It may seem a rather naive notion, but, as we will see shortly, it is far from trivial.

As a simple example, consider the set of left shoes and the set of right shoes that you own. Since each left shoe can be paired with its corresponding right shoe, these two sets are co-numerous. (This assumes that you haven't lost any shoes.) So is the set of husbands and the set of wives (assuming there is no bigamy).

So far, no surprises.

Now consider the set of all numbers from 0 to 2 and the set of all numbers from 0 to 1. We may think of the first set as a line segment 2 inches long and the second as a line segment 1 inch long, as in Figure 4.

0   2

0   1

Figure 4

Are these two sets co-numerous, according to our precise definition? One is twice as long as the other, but length plays no role in the definition of co-numerous. As a matter of fact, these two sets are indeed co-numerous, as Figure 5 suggests.

For each point *P* in the longer segment, draw a line to point *O*. This line,

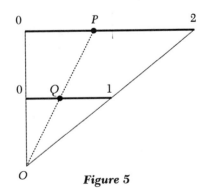

Figure 5

shown dashed in Figure 5, crosses the shorter segment at point Q. In this way, each point in the longer set is paired with exactly one point in the shorter set and vice versa. So, according to the definition of co-numerous, these two sets are co-numerous.

It may seem strange that one segment can be twice as long as another and yet have just as many points as that segment. Though surprising, this shouldn't shock anyone who has ever gone to a movie. The large picture on the screen is just a blowup of a small picture on the film. There is a matching of each point of the large picture with each point of the small one, as shown in Figure 6.

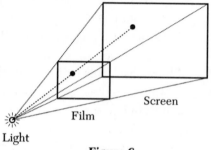

The picture on the screen has a larger area as measured, say, in square inches. But being co-numerous has nothing to do with area. It concerns only being able to pair off the objects in one set with those in another set.

Figure 6

Let's take another example. Let the set N consist of the whole numbers 1, 2, 3, 4, ... The other set, B, consists of the points in the plane that have whole numbers for both coordinates. Parts of these two sets are shown in Figure 7.

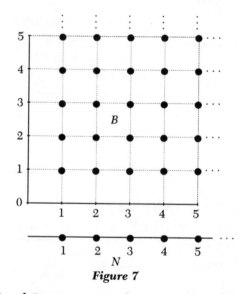

Figure 7

Are the sets N and B co-numerous? Can we pair off each of the numbers 1, 2, 3, 4, . . . with a point in the endless set B, so that every point in

B is paired with a number? At first glance, it may seem impossible. After all, each horizontal row of points in the set B is already co-numerous with the set N. Yet it turns out that the sets B and N are co-numerous. To show that they are, we must describe a way to "marry off" the whole numbers with the points in B.

To accomplish this pairing, imagine that each point in B stands for an orange tree, and you wish to inspect each tree to see if the oranges are ripe. If you simply follow a straight path that stays on the bottom row, you will miss most of the orchard. But there is a route that enables you to inspect every tree. It is suggested by the way a farmer riding a tractor plows a field. It is the zigzag path shown in Figure 8.

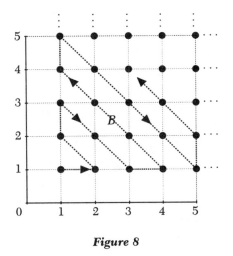

Figure 8

As you walk along this path, starting at the lower left-hand corner, number the trees as you inspect them. This pairs off N, the whole numbers, with the trees in the orchard. Each tree gets tagged by a whole number. Each whole number is the tag of some tree. So the orchard, which at first glance seems "more infinite" than the set of whole numbers, is still co-numerous with it.

Let's stop a moment to look back at what we have done, using the precise notion of being co-numerous. First, we showed that a line segment is co-numerous with one twice as long. Second, we showed that a small rectangle (the film) is co-numerous with a large rectangle (the screen). Third, we showed that an orchard made up of rows, each of which is co-numerous with the set of whole numbers, is co-numerous with that set.

What is going on? Could it be that any two infinite sets are co-numerous? In other words, is there just one "size" of infinity, and is no infinite set "more infinite" than another? For finite sets there certainly are different levels of size. For instance, the set of three bananas is not co-numerous with the set of two apples. No matter how you try to pair the apples with the bananas, one banana will be left over. That's what we mean when we say, "Three is bigger than two," and why we need the two symbols 3 and 2 to name these different levels of being finite.

Let's return to the infinite sets and the question, "Are all infinite sets co-numerous?" This question could have been asked 2,000 years ago, but it wasn't raised until the latter part of the nineteenth century, by Georg Cantor (1845–1918). As he wrote in a letter to another mathematician, Richard Dedekind (1831–1916), on November 29, 1873,

> May I ask you a question . . . which I cannot answer; maybe you can answer it and would be so kind as to write to me about it. It goes as follows. Take the set of all whole numbers and call it N and the set of all positive real numbers and denote it R. Then the question is simply this: can N be paired with R in such a way that to every individual in one set corresponds one and only one individual of the other? At first glance, one says to oneself, "No, this is impossible, for N consists of discrete parts and R is a continuum." But nothing is proved by this objection. And as much as I too feel that N and R do not permit such a pairing, I still cannot find the reason. And it is this reason that bothers me; maybe it is very simple.
>
> Would not one at first glance be led to the conjecture that N cannot be paired off with the set of all positive rational numbers p/q? And yet it is not hard to show that such a pairing can be found.

A pairing of the rationals with the whole numbers that Cantor mentions can be devised with the aid of our orchard. Arrange the fractions like the trees, as in Figure 9. In each row the numerator is kept fixed.

Then follow the same path that you took through the orchard, pairing off whole numbers with fractions as you walk along. Of course, you are even counting ½, ²⁄₄, ³⁄₆, ⅛, . . . as different, though they represent the same rational number. If you just want to count off rational numbers, remove any fractions

⋮	⋮	⋮	⋮	⋮
5/1	5/2	5/3	5/4	5/5 ...
4/1	4/2	4/3	4/4	4/5 ...
3/1	3/2	3/3	3/4	3/5 ...
2/1	2/2	2/3	2/4	2/5 ...
1/1	1/2	1/3	1/4	1/5 ...

Figure 9

that are not reduced to lowest terms from Figure 9, keeping only fractions that are reduced. Then take the same route and count off the remaining fractions as you pass by them in the same zigzag path.

Dedekind replied but did not solve the problem. On December 2, Cantor wrote him again:

> I proposed my question for the following reason. I had asked it several years ago and had always remained in doubt whether the difficulty it presents is subjective or whether it is real, inherent in the substance. I have never seriously thought about it because it has no particular practical interest for me and I fully agree with you if you say that for this reason it doesn't deserve too much labor. Only it would be a beautiful result. . . .

Within a week of writing that letter, Cantor made one of the most dramatic and fundamental discoveries in mathematics: *Not all infinite sets are co-numerous. There are levels of infinity, just as there are levels of finiteness.* The result had an impact on twentieth century mathematics as great as the impact of the discovery of irrational numbers on Greek mathematics 2,300 years earlier.

The argument that Cantor found in 1873 was rather involved, but he found a simpler one in 1890. It is this second one that we shall present.

Recall that N is the set of whole numbers 1, 2, 3, 4, . . . and R is the set of positive real numbers. We will prove that it is impossible to pair off all the members of N with all the members of R. To put it another way, it is impossible to list all the positive real numbers "first," "second," "third," and so on (even though we could, using the zigzag path, list all the positive rational numbers).

What Cantor does is show that for any list of positive real numbers, there is a number that is not on the list. That shows that it is impossible to pair off N with R. Of course, we can pair off N with part of R, say, by pairing 1 with $\frac{1}{1}$, 2 with $\frac{1}{2}$, 3 with $\frac{1}{3}$, and so on, pairing each whole number with its reciprocal. But we will never be able to pair N with all of R.

Now to Cantor's reasoning. Imagine any list whatsoever of positive real numbers. Each number in the list corresponds to one of the numbers 1, 2, 3, ... The list may look, then, like the one at the top of the next page. (I use specific numbers to avoid the clutter of many symbols. The reasoning works for any such list.) Keep in mind that the list is endless and fixed once and for all. The following shows only the first few numbers in the list.

$$1 \ldots \quad 10.387425 \ldots$$
$$2 \ldots \quad 7.084416 \ldots$$
$$3 \ldots \quad 0.250000 \ldots$$
$$4 \ldots 113.333333 \ldots$$
$$5 \ldots \quad 0.912664 \ldots$$

. . . .

Underline the *first* digit to the right of the decimal point in the *first* number in the list. Then underline the *second* digit in the *second* number in the list. Then underline the *third* digit in the *third* number, and so on. These underlinings are shown in the following list.

$$1 \ldots \quad 10.3\underline{8}7425 \ldots$$
$$2 \ldots \quad 7.0\underline{8}4416 \ldots$$
$$3 \ldots \quad 0.25\underline{0}000 \ldots$$
$$4 \ldots 113.333\underline{3}33 \ldots$$
$$5 \ldots \quad 0.9126\underline{6}4 \ldots$$

. . . .

Pay no attention to the digits that are not underlined. They won't play any role in the reasoning. Only the string of underlined digits matter, those that form an endless downward diagonal.

Cantor uses that diagonal to construct a number, call it r, that he shows cannot be on the list. The number r will look like this:

$$r = 0.\_ \_ \_ \_ \_ \cdots$$

We have to tell what digits go in the waiting, empty places.

For convenience, write the underlined digits in order:

$$\underline{3}\ \underline{8}\ \underline{0}\ \underline{3}\ \underline{6} \ldots$$

Below each one of these digits write a digit that is different from it. To do this automatically, without having to make a choice each time, we make up a rule: If the underlined digit is 8, write a 7; if the underlined digit is not 8, then write an 8. Then r is the number whose decimal form is given by our rule. In our case it begins

$$r = 0.87888 \ldots$$

This number r cannot appear anywhere in the given list. (Think about this before reading on. Could it be the first number? the second? the third?)

It cannot be the first number because it differs from the first number already at the very first place after the decimal point. It cannot be the second number because it differs from that number already at the second place. It cannot be the third number because it differs from it already at the third place. More generally, it differs from the nth number in the list at least in the nth decimal place. So r cannot appear anywhere on the list. This argument applies to all possible lists, not just the one I started.

We must conclude that it is impossible to pair off N with R. The level of infinity represented by the set of positive real numbers is truly larger than the level of infinity represented by the set of whole numbers. Just as when you try to pair off two apples with three bananas, you will have a banana left over, when you try to pair off the whole numbers with the reals, you will always have reals left over.

We must conclude that there are various "sizes" of infinite sets. The smallest size is represented by the set of whole numbers. Any set that can be paired off with this set is called *denumerable*. We showed that the rationals are denumerable, but the reals are not.

Are the irrational numbers denumerable? If they were, we could pair them off with the (endless) lower line of dots in Figure 10.

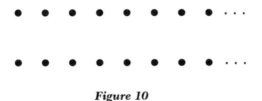

Figure 10

Then we could pair all the rational numbers with the upper line of dots. In that way, we would have paired all the real numbers with all the dots in Figure 10. But the same method that showed that the orchard in Figure 8 is denumerable shows that the dots in Figure 10 are denumerable. (The farmer could inspect all the trees in just two rows.) That implies that all the real numbers would be denumerable. Since they are not, we must conclude that the irrationals are not denumerable. That means that there are more irrational numbers than rational numbers.

The symbol for being denumerable is $\aleph_0$, pronounced "aleph null." Just as our set of two apples has 2 members, the set of whole numbers has $\aleph_0$ members. The number $\aleph_0$ is the first one after all the whole numbers.

Incidentally, the symbol ℵ is the first letter of the Hebrew alphabet. Cantor chose it because it stands for the number one (the unit) in Hebrew, and he thought of an infinite set as having a unity, or "oneness." That is why, over a century later, ℵ (with various subscripts) is used in naming the sizes of infinity. (The symbol for the size of the set of real numbers is c, for *continuum*.)

Cantor's discovery is not only "a beautiful result." It has had a profound impact on mathematics and logic. For instance, in advanced calculus, topology, and algebra, careful attention must be paid to the size of the infinite sets. In these fields, there are many theorems that hold for denumerable sets that do not hold for all infinite sets. Moreover, Cantor's *diagonal argument* appears even in logic and the theory of computing machines.

The question Cantor posed, seemingly so naive and unimportant, started a revolution. In mathematics, as in many disciplines, asking the right question is every bit as important as finding the right answer.

PART THREE

Closer and Closer

··· 27 ···

Zero over Zero

Imagine Socrates asking a question of his two young friends, Patrick and Patricia.

Socrates: What happens to the quotient

$$\frac{x^2 - 1}{x - 1}$$

when x gets near 1?

Patrick: That's easy. When x is 1, the numerator is $1^2 - 1$, which is 0. The denominator is $1 - 1$, which also is 0. So the quotient becomes

$$\frac{0}{0}.$$

Since anything divided by itself is 1,

$$\frac{x^2 - 1}{x - 1}$$

approaches 1 as x approaches 1.

Patricia: That doesn't make any sense.

Socrates: Why not?

Patricia: Division by zero is ridiculous.

Socrates: Why?

Patricia: Division is a way of looking at multiplication. When we say "6 divided by 2 is 3," we are answering the multiplication question

"2 times what is 6?" In other words, we are filling the box in the equation

$$2 \times \square = 6.$$

There is one way, just one way, to fill that box. But to talk about "0 divided by 0," we have to fill the box in

$$0 \times \square = 0.$$

The trouble is that any number can go in the box. For instance,

$$0 \times 5 = 0$$

and

$$0 \times 7 = 0.$$

So it makes no sense to talk about "0 divided by 0." It's not at all like "6 divided by 2."

Socrates: But Patrick said that anything divided by itself is 1.

Patricia: He's right, except for the number 0.

Socrates: So what about my question? What does happen to

$$\frac{x^2 - 1}{x - 1}$$

when x gets near 1?

Patrick: I still think it approaches 1.

Socrates: How can we find out?

Patricia: Just try some choices for x near 1 and see what happens.

Socrates: What number do you want to use?

Patricia: As a starter, let's try 1.1.

Patrick: I'll do the arithmetic:

$$\frac{(1.1)^2 - 1}{1.1 - 1} = \frac{1.21 - 1}{0.1} = \frac{0.21}{0.1} = 2.1.$$

Well, it's not near 1. I still think that if we choose x nearer to 1, then your quotient will be near 1.

Socrates: Go ahead.

Patrick: I'll try $x = 1.01$. That should settle the matter. Now the quotient is

$$\frac{(1.01)^2 - 1}{1.01 - 1} = \frac{1.0201 - 1}{0.01} = \frac{0.0201}{0.01} = 2.01.$$

Uh-oh, it's not near 1. I'll change my opinion. I feel that the quotient will approach 2 as x gets near 1.

Socrates: Why 2?

Patrick: That's the only famous number near 2.01.

Socrates: Why must the quotient approach a famous number?

Patrick: Otherwise I'd have to give up.

Patricia: Couldn't we try some number even closer to 1 and see what happens?

Socrates: Go ahead.

Patricia: I'll try $x = 1.0001$. Then I get

$$\frac{x^2 - 1}{x - 1} = \frac{1.0001^2 - 1}{1.0001 - 1} = \frac{1.00020001 - 1}{0.0001} = \frac{0.00020001}{0.0001} = 2.0001.$$

That sure is close to 2. Maybe Patrick is right this time. I think that the quotient approaches 2 as x gets near 1.

Socrates: Are you absolutely certain?

Patricia: Yes.

Socrates: But maybe the quotient approaches 1.999999976? Isn't that still possible?

Patrick: It's possible, but I would be surprised if it does.

Socrates: At this point you both agree and you have a strong opinion. But you haven't really settled the issue beyond a shadow of doubt.

Patricia: I have an idea. As long as we use specific numbers, we'll never be absolutely sure. But that quotient

$$\frac{x^2 - 1}{x - 1}$$

reminds me of some quotients I've seen before, only the letter was r instead of x. I remember. There was this equation

$$\frac{1 - r^{k+1}}{1 - r} = 1 + r + r^2 + \ldots + r^k.$$

I saw it in this very book.
The case $k = 1$ gives

$$\frac{1 - r^2}{1 - r} = 1 + r.$$

But

$$\frac{1 - r^2}{1 - r} = \frac{r^2 - 1}{r - 1}.$$

So

$$\frac{r^2 - 1}{r - 1} = 1 + r.$$

This equation makes sense whenever r isn't 1. When I use the letter x instead of r, the left side is exactly your quotient:

$$\frac{x^2 - 1}{x - 1} = 1 + x.$$

Patrick: So what?

Patricia: It's a lot easier to see what happens to $1 + x$ than what happens to the quotient.

Patrick: You're right. When x is near 1, then $1 + x$ is near $1 + 1$, which is 2. I don't have to worry about division by zero. So that quotient does approach 2 as x gets near 1. That's what I said all along.

Socrates: So what happens to

$$\frac{x^3 - 1}{x - 1}$$

as x gets near 1?

Patrick: I bet it gets near 2.

Patricia: I don't know, but I'll use the same method on this, and write

$$\frac{x^3 - 1}{x - 1} = 1 + x + x^2.$$

As x gets near 1, then $1 + x + x^2$ gets near $1 + 1 + 1^2$, which is 3. So I'm sure that the quotient approaches 3.

Socrates: So Patrick's guess is wrong?

Patrick: This isn't fair. She used a trick.

Socrates: The first time she used it, you could call it a trick. But not the second time. By then it's just a tool, like a hammer or pliers. Do you call a hammer a trick?

Patrick: An equation is not a hammer. I still think that the quotient approaches 2.

Socrates: A stubborn fellow.

Patricia: If he isn't convinced already that it approaches 3, I think I can bring him around.

Socrates: How?

Patricia: I'll use a specific number near 1, say 1.01. Then the quotient is

$$\frac{1.01^3 - 1}{1.01 - 1} = \frac{1.030301 - 1}{0.01} = 3.0301.$$

Well, Patrick?

Patrick: I give up, it's 3.

Socrates: What is the moral of our conversation?

Patrick: Stay away from zero over zero.

Patricia: No. It's be careful when you see zero over zero.

Socrates: Yes. When you divide a small number by a small number, the quotient might be gigantic or tiny or anywhere in between. That's not so with the product of two small numbers, or their sum, or their difference. In each case you get a small number. Division is another story entirely. In the division of one small number by another, anything can happen.

··· 28 ···

How Steep Is a Curve?

Looking at a map of San Francisco, you'd never suspect that there were any hills in the older part of the city. The streets there follow straight lines. Instead of going along the contours, they ruthlessly climb up and down the hills, undaunted by any obstacle. There was a brief attempt after the 1906 earthquake to have them acknowledge nature, but it failed, for the people moved right back to their old lots. When you drive up one of the steep streets and are about to reach the top, you see only blue sky in front of you. You are not sure whether the street goes on or you are about to fall off a cliff.

How steep is the steepest street in San Francisco? People usually guess that it would make an angle of about 45° with the horizontal. That means a rise of 1 foot for every foot of run, that is, of horizontal motion. That steepness is shown in Figure 1.

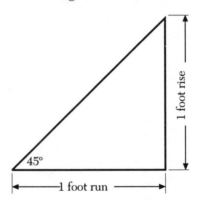

Figure 1

As a matter of fact, the steepest street used by automobiles is Filbert, and its angle is a mere 17.5°, a rise of only 31.5 feet in a run of 100 feet. Its steepness is shown in Figure 2.

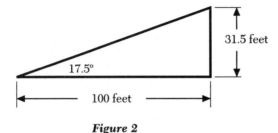

Figure 2

As we see, there are two ways to measure the steepness of a line. Either give its angle with the horizontal or tell how much it rises for a given (horizontal) run. We will use the second method, describing the steepness by the quotient

$$\frac{\text{rise}}{\text{run}}.$$

Call this quotient the *slope* of the line. Carpenters call it the *pitch* (of a roof), and highway engineers call it the *grade* (of a road). By the way, the minimum pitch of an A-frame house should be 1.3 in order to shed snow. The grade of any part of the interstate highway system is at most 0.06. The steepest grade permitted for a new road in California is 0.09, allowable only in an urban hilly area.

The slope of the slanted line in Figure 1 is ¼, or 1. In Figure 2, it is ³¹·⁵⁄₁₀₀ = 0.315. The bigger the slope, the steeper the line.

Now let us consider the slopes of lines in the *xy* plane instead of the slopes of hills. Let *L* be a line in the *xy* plane, as shown in Figure 3.

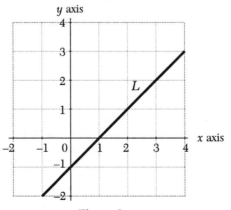

Figure 3

Pick any two points on L, and call them P and Q, with P to the left of Q, as in Figure 4.

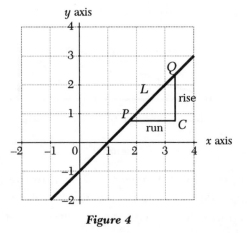

Figure 4

These points determine a right triangle PCQ with sides parallel to the axes, as in Figure 4. The length of PC is the run. It is the change in x. The vertical side QC determines the rise. It is the change in y. If, as we move from P to Q on the line, we move upward, as in Figure 4, the rise is taken to be positive. However, if, as we move from left to right on the line, we go downward, the rise is taken to be negative. This is the case with the line shown in Figure 5. (The phrase *a negative rise* sounds odd, but so does the phrase *negative growth,* which describes a shrinking economy or population. Besides, *elevators* go down as well as up.)

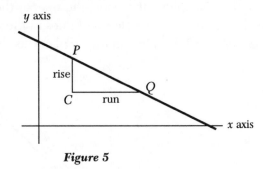

Figure 5

The steepness, or *slope,* of a line in the plane is the quotient

$$\frac{\text{rise}}{\text{run}},$$

which can be negative, as in Figure 5. (Slopes of roads, roofs, and so on are always taken to be positive.)

For practice, draw x and y axes and a line L not parallel to the y axis. Pick a couple of points, P and Q, on the line. Measure the run and rise they define (preferably in centimeters, for ease of arithmetic). Then calculate the slope. Do this for another choice of two points on the line. You should get the same slope—or almost the same, since there are always little errors in the measurements.

So much for the slope of a line. But what about the slope of a curve? Figure 6 shows part of the curve $y = x^2$.

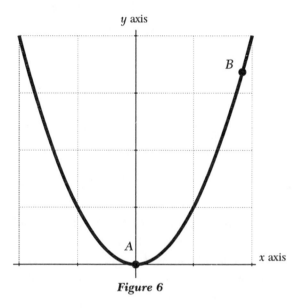

Figure 6

The part near point A is not at all steep. As you move along the curve from A to the right, the steepness increases. At point B, the curve is very steep. How can we possibly measure the steepness of a curve, when the steepness changes from point to point?

To be specific and down to earth, say we want to find how steep the curve is at the point $P = (1, 1)$. We could use a ruler to draw a line T that goes through P and seems to be moving in the same direction as the curve does there. T just touches the curve and does not cross it. Call T a *tangent* to the curve, in honor of the Latin *tangere* (to touch).

There will naturally be some error in drawing T, since it's hard to estimate the "direction" of the curve near P. But at least we could estimate the slope of T by using two points on it, say P and some other point, Q. This is done in Figure 7.

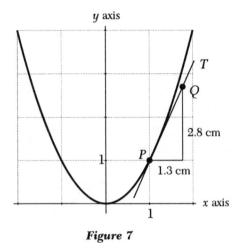

Figure 7

In my estimate, the run is about 1.3 centimeters and the rise is about 2.8 centimeters. So my estimate of the slope is

$$\text{slope} \approx \frac{2.8}{1.3} \approx 2.2.$$

This is just an estimate. I suggest that you make a much larger sketch of the curve $y = x^2$ and draw a tangent line T at $P = (1, 1)$ as well as you can. Then make a large triangle along T to get a rise and run.

If you had enough time and paper, you might make a gigantic picture of the curve, draw a tangent at $(1, 1)$, and estimate the slope a bit more accurately. But you will notice that just a slight change in the angle of the line T can cause a big change in the slope. Besides, an estimate is just an estimate, and even a hundred estimates are just that—a hundred estimates.

How, then, can we find the exact slope of the tangent at $(1, 1)$? Keep in mind that the slope of this tangent, T, will be the same as the slope of the curve at $(1, 1)$. Fortunately, there is another way to approach this challenge, which doesn't require that we draw anything. Let's now look at this other way.

Choose a point Q on the curve fairly near P. The points P and Q determine a line L. This line is called the *secant* determined by P and Q. The part of this line that lies between P and Q is called the *chord* determined by P and Q. L is certainly not the tangent at P, but it does resemble it somewhat. This is shown in Figure 8.

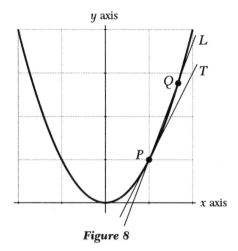

Figure 8

The closer you choose Q to P, the more line L looks like the tangent T. You may check this yourself on your own diagram. So the slope of L provides an estimate of the slope of T.

Just to get a feel for this different approach, let's see what estimates we obtain for some specific choices of the point Q. For instance, choose Q to be the point on the curve whose x coordinate is 1.1. Since the curve has the equation $y = x^2$, the y coordinate of Q is the square of 1.1, that is, 1.21. Figure 9 shows $P = (1, 1)$ and $Q = (1.1, 1.21)$.

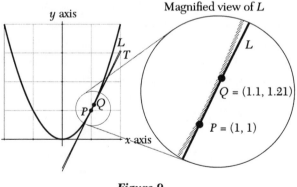

Figure 9

To find the slope of the line L through P and Q, draw the little "rise and run" triangle determined by those two points. Figure 10 shows this triangle, somewhat magnified.

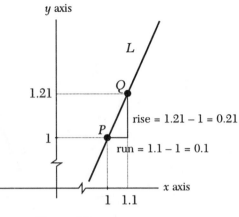

Figure 10

The rise in the little triangle is exactly $1.21 - 1 = 0.21$, and the run is $1.1 - 1 = 0.1$. The slope of the approximating line L through P and Q is therefore

$$\frac{\text{rise}}{\text{run}} = \frac{0.21}{0.1} = 2.1.$$

This is just another estimate of the slope of the tangent line at P. But, in contrast with our previous estimate, we really didn't have to draw anything at all. Figure 10 simply served as a guide to our thinking. Our calculations did not depend on it whatsoever. Of course, it is just an estimate.

To get a better estimate of the slope at P, choose a point Q on the curve even closer to $P = (1, 1)$. For instance, this time choose the point Q to have an x coordinate 1.01. So $Q = (1.01, 1.01^2) = (1.01, 1.0201)$. Without making any drawings at all, we find the slope of the line PQ to be

$$\frac{\text{rise}}{\text{run}} = \frac{1.0201 - 1}{1.01 - 1} = \frac{0.0201}{0.01} = 2.01.$$

Notice the magic: We didn't even draw the curve or P or Q or the line through them.

We still don't know the slope of the tangent T at P. Our last two estimates, 2.1 and 2.01, may suggest that the slope of T is 2. That is the only famous number near 2.01, but we can't be sure that the slope of T is a famous number. For all we know, it might be 1.987 or some complicated expression involving the square root of 2 or something even fancier.

We have to find out exactly what happens to the slope of the line L

through P and Q as Q approaches P. In order to do this, we must stop choosing specific points Q and consider all possible Q at one blow.

To do this, let x be any number larger than 1 and let Q be the point (x, x^2). Let's see what happens to the slope of the line through $P = (1, 1)$ and $Q = (x, x^2)$ as x gets nearer and nearer 1. Figure 11 shows the general case.

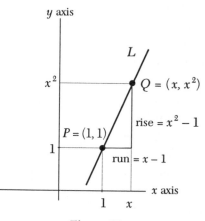

Figure 11

In this general case, the slope of the approximating line is

$$\frac{\text{rise}}{\text{run}} = \frac{x^2 - 1}{x - 1}. \qquad (1)$$

We have to find out what happens to the quotient (1) as x gets nearer and nearer 1. But, by a happy coincidence, we already found out how this quotient behaves as x approaches 1. In Chapter 27, after showing that the quotient equals $x + 1$, we saw that it approaches 2.

The quotient

$$\frac{x^2 - 1}{x - 1}$$

approaches the slope of the tangent line T to the curve at $(1, 1)$ and also the number 2. So the slope of the tangent line is indeed a famous number, namely, 2. And we found it without having to draw anything at all.

What do we mean, then, by "the slope of the curve $y = x^2$ at $(1, 1)$?" We mean the slope of the tangent line to the curve at $(1, 1)$, for that line suggests the direction of the curve near $(1, 1)$. We now can say that the slope of the curve at $(1, 1)$ is exactly 2.

To check that you understand the *nearby point* technique, I suggest that you work with the curve $y = x^3$ instead, and find the slope of the tangent line at the point (1, 1), which lies on the curve. (You will meet the quotient $(x^3 - 1)/(x - 1)$.) Then do the same thing for the curve $y = x^4$.

The nearby point method works not just at the point (1, 1), but at any point on these curves. It just requires a little more algebra in the spirit of Chapter 27. The essential idea of this technique is already displayed in the cases we considered. In a calculus course, the technique is applied to a great variety of curves.

The slope of a curve plays an important role in economics, biology, engineering, physics, chemistry, business administration, and so on. One reason is that if we know the formula for the slope, we can then find the highest and lowest points on a curve. To see why, look at the curve in Figure 12.

Imagine drawing the tangent line to the curve at *H*, the high point, or at *B*, the low point. These

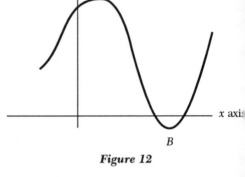

Figure 12

lines, being parallel to the *x* axis, have a slope of 0, as indicated in Figure 13. (Imagine a ruler that you keep parallel to the *x* axis. Then slide it up until it just touches the curve or down until it just touches the curve.)

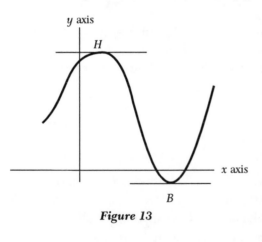

Figure 13

If we had a formula for the slope of the curve at all its points, we would look for those points at which the slope is 0. That means solving an equation, a skill developed in algebra classes. A businessperson who wants to maximize profit or an engineer who wants to design an economical package will be looking for high or low points on curves. That is just one of the many ways that calculus serves as a tool. Some of the other ways are described at the end of Chapter 30.

··· 29 ···

Trying to Find a Curved Area

In school we learned to find the areas of certain figures. For instance, the area of a rectangle of sides a and b is their product ab, as shown in Figure 1.

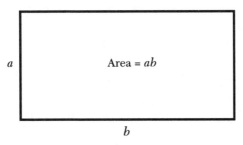

Area = ab

Figure 1

With the aid of this fact we found the area of a parallelogram of base b and height h, as shown in Figure 2.

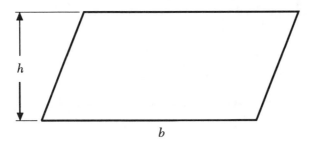

Figure 2

Just cut off the shaded triangle in Figure 3 and slide it to the left, changing the parallelogram into a rectangle of the same area.

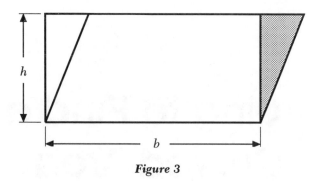

Figure 3

Since the rectangle has base b and height h, its area is bh. Therefore, the area of the parallelogram is also bh.

It is now a quick step to find the area of a triangle of base b and height h, as shown in Figure 4.

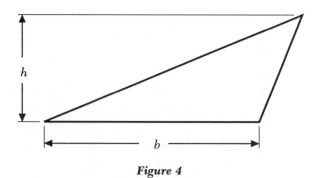

Figure 4

Make an exact copy of this triangle, spin it around, and put it next to the original triangle, as in Figure 5, to form a parallelogram.

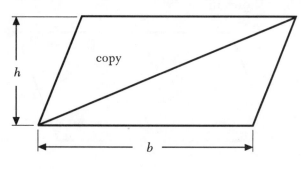

Figure 5

Since the area of the parallelogram formed is bh, the area of each triangle is $(\frac{1}{2})bh$, that is, half the base times the height.

Since we can find the area of any triangle, we can find the area within any polygon: Simply cut it into triangles and find the area of each triangle.

But how do we find the area of a region bounded by a curve rather than by straight segments? For instance, how do we find the area below the curve $y = x^2$ and above the section where x goes from 0 to 1? That area is shaded in Figure 6. (Looking at Figure 6, we can say that the area is less than $\frac{1}{2}$, since it lies inside a triangle whose base is 1 and whose height is 1.)

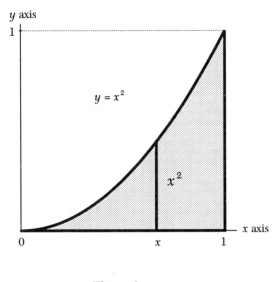

Figure 6

One way is to build an approximating staircase made up of thin rectangles, a method that goes back to Archimedes in the third century B.C.

In this approach, you first pick a positive integer n and chop the interval from 0 to 1 into n pieces of equal length, using $n - 1$ equally spaced points. Then you build a rectangle on each piece. The height of each rectangle equals the height of the curve $y = x^2$ where it meets the right-hand edge of the rectangle.

To bring things down to earth, take the case $n = 5$. Break the interval from 0 to 1 into five equal pieces, as in Figure 7. One of the five rectangles is shown in that figure.

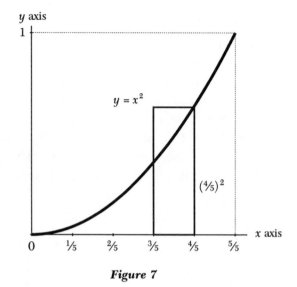

Figure 7

The whole staircase for the case $n = 5$ is shown in Figure 8.

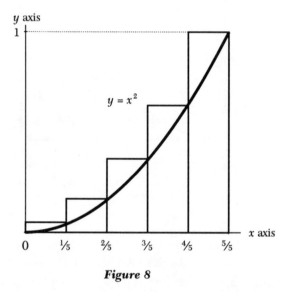

Figure 8

The base of each rectangle in Figure 8 is ⅕. The height of the smallest one is $(⅕)^2$, since the curve has the equation $y = x^2$. (When x is ⅕, then y is $(⅕)^2$.) So the area of this smallest rectangle, being the product of its height and its base, is

$$\left(\frac{1}{5}\right)^2\left(\frac{1}{5}\right) = \frac{1}{5^2} \times \frac{1}{5} = \frac{1^2}{5^3}.$$

The height of the rectangle just to its right is $(\frac{2}{5})^2$. So its area is

$$\left(\frac{2}{5}\right)^2\left(\frac{1}{5}\right) = \frac{2^2}{5^2} \times \frac{1}{5} = \frac{2^2}{5^3}.$$

Note that the denominator is 5^3 for each rectangle. Only the numerator changes.

The areas of the other three rectangles are found the same way. The total area of the five rectangles is

$$\frac{1^2}{5^3} + \frac{2^2}{5^3} + \frac{3^2}{5^3} + \frac{4^2}{5^3} + \frac{5^2}{5^3}.$$

This sum can be written a little more simply by using the common denominator 5^3, as

$$\frac{1^2 + 2^2 + 3^2 + 4^2 + 5^2}{5^3}. \qquad (1)$$

The fraction labeled (1) is only an approximation to the area under $y = x^2$, based on five rectangles. (It equals $\frac{55}{125}$, which is 0.44.)

If we instead use $n = 10$, say, we will obtain a staircase made of ten rectangles, as in Figure 9. The area of this staircase is a better approximation to the area under the curve.

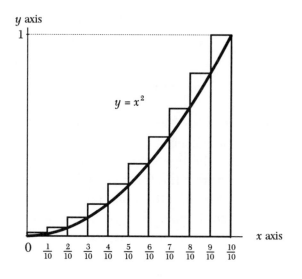

Figure 9

The total area of this staircase is, by a similar calculation,

$$\frac{1^2 + 2^2 + 3^2 + 4^2 + 5^2 + 6^2 + 7^2 + 8^2 + 9^2 + 10^2}{10^3}.$$

(This estimate is 0.385.)

The staircases formed for $n = 5$ and $n = 10$ are typical of any value of n, which is the general case. Figure 10 indicates the general case, where each of n rectangles has the width $\frac{1}{n}$.

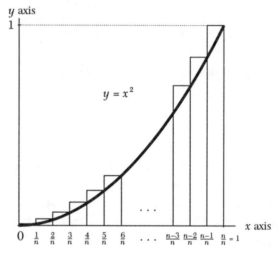

Figure 10

The area of this typical staircase is

$$\frac{1^2 + 2^2 + 3^2 + \ldots + n^2}{n^3}. \tag{2}$$

What happens to the fraction (2) as n increases? If we knew, then we would know the area under the curve. As n grows, so does the numerator. That might suggest that the fraction gets large. But the denominator also gets large. That suggests that the fraction might get small. If the denominator grows much faster than the numerator, the fraction might even approach 0. That cannot happen, however, since it approaches the area under the curve, a number that is bigger than 0 and, as we noted, less than $\frac{1}{2}$.

Clearly, there is a battle being waged between the numerator and the denominator. What should we do? We could compute the value of (2) for a variety of choices of n and then make a guess. For $n = 5$, we found that the fraction is 0.44, and for $n = 10$, it is 0.385. It looks as though the values decrease as n increases. Moreover, these estimates are always larger than the area under the curve, since the staircases are higher than the curve.

If you have a calculator handy, try $n = 11$, 12, and so on, just to get a feel for the way the fraction behaves. Maybe your experiments will suggest what the area under the curve is. Maybe not. In any case, we are blocked because we don't know what number the fractions approach. In Chapter 30, using a different shape of staircase, we do find the area under the curve. So, indirectly, we find what happens to fraction (2)—information that is useful in Chapter 31. That means that our work in this chapter, though it did not find the area, has not been in vain.

In Chapter 31, we also need to know what happens to other similar fractions as n increases. These fractions are formed with larger exponents, such as the fraction

$$\frac{1^4 + 2^4 + 3^4 + \ldots + n^4}{n^5}. \tag{3}$$

This fraction differs from (2) only in the replacing of the exponent 2 in the numerator by the exponent 4, and the replacing of the exponent 3 in the denominator by the exponent 5. As you may check by drawing a typical staircase, it approximates the area under the curve $y = x^4$ from 0 to 1. I suggest that you play with it for various values of n, and guess what number it approaches as n increases. (In the next chapter, we find that number.)

··· 30 ···

Finding a
Curved Area

In Chapter 29, we tried to find the area under the curve $y = x^2$ but failed. In our technique, the rectangles we used to make a staircase all had the same width. But there is another method, due to Fermat's work in the seventeenth century, which does get that area exactly. In his method, the rectangles do not have the same width.

We will follow Fermat's reasoning, which uses two facts obtained in Chapter 18.

The first is that for any number r between -1 and 1,

$$1 + r + r^2 + r^3 + \ldots = \frac{1}{1 - r}.$$

He needs this equation when r is the cube of another number, p; that is, $r = p^3$. So he uses the equation

$$1 + p^3 + (p^3)^2 + (p^3)^3 + \ldots = \frac{1}{1 - p^3}.$$

This tells him that

$$1 + p^3 + p^6 + p^9 + \ldots = \frac{1}{1 - p^3}.$$

The second is that

$$1 + x + x^2 + \ldots + x^k = \frac{1 - x^{k+1}}{1 - x},$$

for any whole number k and any number x other than 1. (We use the letter x rather than the letter r to avoid confusion later in the chapter.) Actually, he uses this second fact upside down and in the reverse order, rewriting it as

$$\frac{1-x}{1-x^{k+1}} = \frac{1}{1+x+x^2+\ldots+x^k}.$$

For each positive number p less than 1, he makes a corresponding staircase, as follows. First he marks off where the numbers p, p^2, p^3, p^4, ... are located on the x axis, as in Figure 1. As the exponent k increases, p^k decreases, getting closer to 0.

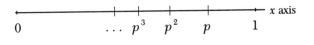

Figure 1

As we move from right to left, the points chosen get closer and closer together. They also get closer and closer to 0.

Next, he computes the height of the curve $y = x^2$ above each of these points. For instance, when $x = p$, then $y = x^2 = p^2$. When $x = p^2$, then $y = x^2 = (p^2)^2 = p^4$. When $x = p^3$, then $y = (p^3)^2 = p^6$. Some of these heights are shown in Figure 2.

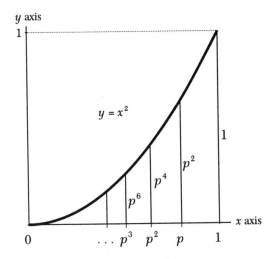

Figure 2

Then he makes a staircase of rectangles, with the height of each rectangle equal to the height of the curve at the right-hand edge of the rectangle, as in Figure 3. (In this diagram, p is 0.9. You might try $p = 0.99$ and draw a few of the rectangles in this case.)

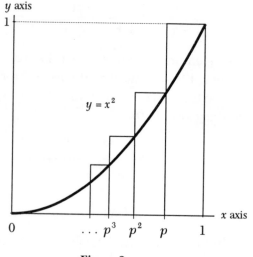

Figure 3

All that remains is to find the total area of this staircase and then see what happens to this area as p gets nearer and nearer to 1. (The nearer p is to 1, the narrower the rectangles and the more the staircase resembles the curve.) This is a matter of careful bookkeeping, that is, finding the area of each rectangle and then totaling up all those areas.

Consider first the area of the largest rectangle. Its height is 1 and its width is $1 - p$. Its area is therefore

$$1 \times (1 - p).$$

Next look at the rectangle just to its left. Its height is p^2 and its width is $p - p^2$. Its area is therefore

$$p^2(p - p^2) = p^2 p(1 - p) = p^3(1 - p).$$

Next consider the third rectangle from the right. Its height is p^4 and its width is $p^2 - p^3$. Its area is

$$p^4(p^2 - p^3) = p^4 p^2(1 - p) = p^6(1 - p).$$

The area of the next rectangle, as you can check, is

$$p^9(1 - p).$$

The area of each rectangle is the product of $1 - p$ and a power of p. The exponent in this power increases by three each time we move to the left in the staircase of rectangles. The area of the endless staircase is therefore

$$(1-p) + p^3(1-p) + p^6(1-p) + p^9(1-p) + \ldots .$$

Since $1-p$ appears in each term, factor it out, obtaining

$$(1-p)(1 + p^3 + p^6 + p^9 + \ldots).$$

The endless sum that appears in this equation is a geometric series in disguise, with ratio p^3. As mentioned at the beginning of the chapter, its sum is $1/(1-p^3)$. Therefore,

$$\text{Area of staircase} = (1-p)(1 + p^3 + p^6 + p^9 + \ldots) = (1-p) \times \frac{1}{1-p^3}.$$

Fermat finally has a very short formula for the area of the staircase:

$$\text{Area of staircase} = \frac{1-p}{1-p^3}.$$

How satisfying that after so much work, we reach such a simple formula. All that is left is to find out what happens to

$$\frac{1-p}{1-p^3}$$

as p gets nearer and nearer to 1.

At the beginning of the chapter, we mentioned that for any whole number k,

$$\frac{1-x}{1-x^{k+1}} = \frac{1}{1 + x + x^2 + \ldots + x^k}.$$

That is exactly what Fermat needs right now in the special case that k is 2 and x is now p. He has

$$\text{Area of staircase} = \frac{1-p}{1-p^3} = \frac{1}{1 + p + p^2}.$$

As p gets nearer and nearer to 1, the area of the staircase approaches

$$\frac{1}{1 + 1 + 1^2} = \frac{1}{1 + 1 + 1} = \frac{1}{3}.$$

The area under the curve must be ⅓.

Is this answer reasonable? A glance at Figure 4 shows that it is. The curved area that we have found lies inside a triangle whose base is 1 and whose height is also 1; hence, it has area $\frac{1}{2}$. The curved area occupies exactly $\frac{2}{3}$ of the area of this triangle, since $\frac{1}{3}$ is equal to $\frac{2}{3} \times \frac{1}{2}$.

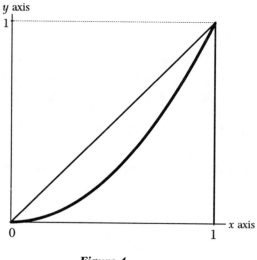

Figure 4

A similar approach works for the curve $y = x^3$. As you follow the steps, you will find that you will need two pieces of information:

$$1 + p^4 + p^8 + \ldots = \frac{1}{1 - p^4}$$

and

$$\frac{1 - p}{1 - p^4} = \frac{1}{1 + p + p^2 + p^3}.$$

Both of these are just special cases of equations collected at the beginning of the chapter. As p approaches 1, therefore, the area of the staircase approaches

$$\frac{1}{1 + 1 + 1^2 + 1^3} = \frac{1}{1 + 1 + 1 + 1} = \frac{1}{4}.$$

We conclude that the area under the curve $y = x^3$ for x between 0 and 1 is $\frac{1}{4}$. If you graph the curve, you will see that it lies below the curve $y = x^2$ when x is between 0 and 1. So it is to be expected that the area beneath it is less than $\frac{1}{3}$, our answer in the case of $y = x^2$.

The same argument, followed step by step, works for all the curves $y = x^4, y = x^5, y = x^6, \ldots$. You can check that for any whole number k, the area under the curve $y = x^k$ and above the interval from 0 to 1 is $1/(k + 1)$.

Looking back at Fermat's argument reveals its simplicity: approximate a curved area by an area that we can compute exactly. Then see what happens to these approximations as they get better and better.

Now that we have found the area under the curve $y = x^2$, we know what happens to the quotient we met in Chapter 29,

$$\frac{1^2 + 2^2 + \ldots + n^2}{n^3},$$

as n increases. Since it approaches that area, the quotient must approach ⅓.

Once we know that $(1^2 + 2^2 + \ldots + n^2)/n^3$ approaches ⅓ as n increases, we are free to apply this fact wherever it comes in handy. For instance, we can use it to show that the volume of a ball of radius r is $4\pi r^3/3$. I will just sketch the reasoning in the case when the radius is 1, and leave the details to you.

Consider the volume of just a hemisphere. (This will simplify the arithmetic. The volume of the ball will be twice as large.) Approximate the hemisphere by a stack of n "coins." To do this, pick a whole number n and divide the radius that is perpendicular to the base of the hemisphere into n sections, each of the length ⅟ₙ. Then use n coins, each of thickness ⅟ₙ, as in Figure 5, which shows the case $n = 5$. (Note that the smallest coin has radius 0 and the coins are inside the ball.)

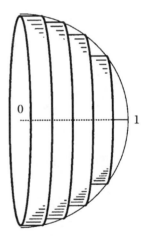

Figure 5

Compute the total volume of the coins. (You will need the Pythagorean theorem to find the radius of each coin.) As a check on your calculations, here is the estimate for the case when n is 5:

$$\pi\left(\frac{1}{5}\right)\left\{\left[1-\left(\frac{1}{5}\right)^2\right]+\left[1-\left(\frac{2}{5}\right)^2\right]+\left[1-\left(\frac{3}{5}\right)^2\right]+\left[1-\left(\frac{4}{5}\right)^2\right]+\left[1-\left(\frac{5}{5}\right)^2\right]\right\},$$

which equals

$$\pi\left(\frac{1}{5}\right)\left(5 - \frac{1^2 + 2^2 + 3^2 + 4^2 + 5^2}{5^2}\right).$$

Then find what happens to your sums as n increases. The same approach goes through for the general case, when the radius r need not be 1.

Once you know the formula for the volume of any ball, it is a short step to find the area of the surface of a ball. Again I will sketch the reasoning only in the case when the radius is 1.

Imagine two balls with the same center, one of radius 1 and one of radius s, which is a bit larger than 1. Draw a picture of the thin spherical shell that lies inside the larger ball and outside the smaller ball. The volume of this shell is

$$\frac{4\pi s^3}{3} - \frac{4\pi 1^3}{3},$$

which equals

$$\frac{4\pi(s^3 - 1)}{3}.$$

Let the surface area of the inner ball be A. The volume of the shell is roughly A times the thickness of the shell, $s - 1$. So the area A is approximated by the quotient

$$\frac{4\pi(s^3 - 1)/3}{s - 1}.$$

All that remains is to find what happens to this quotient as s approaches 1. The equation $s^3 - 1 = (s - 1)(s^2 + s + 1)$—which is just the formula for the sum of the geometric series $1 + s + s^2$ in disguise—will help here, since it will enable you to get rid of the pesky $s - 1$ in the denominator.

Using the area under $y = x^k$ shows that

$$\frac{1^k + 2^k + \ldots + n^k}{n^{k+1}}$$

approaches $1/(k + 1)$ as n increases. You will need this information in the next chapter.

As we compare this chapter with the one before, we see that the method tried in the other chapter failed because we didn't have a short formula for the sum $1^k + 2^k + 3^k + \ldots + n^k$. In fact, there are such formulas, and, when k is small, they can be found with some effort. Just for the record, here are the formulas for the first four cases, $k = 1, 2, 3,$ and 4:

$$1 + 2 + 3 + \ldots + n = \frac{n^2}{2} + \frac{n}{2}$$

$$1^2 + 2^2 + 3^2 + \ldots + n^2 = \frac{n^3}{3} + \frac{n^2}{2} + \frac{n}{6}$$

$$1^3 + 2^3 + 3^3 + \ldots + n^3 = \frac{n^4}{4} + \frac{n^3}{2} + \frac{n^2}{4}$$

$$1^4 + 2^4 + 3^4 + \ldots + n^4 = \frac{n^5}{5} + \frac{n^4}{2} + \frac{n^3}{3} - \frac{n}{30}.$$

You can get the first of these formulas by looking at the following picture of an n by $n + 1$ rectangle, broken into two identical staircases.

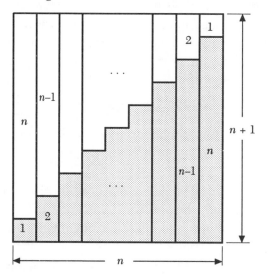

The area of each staircase is $1 + 2 + 3 + \ldots + n$. Since the area of the rectangle is $n(n + 1)$, the area of each staircase is half of $n(n + 1)$. That tells us that

$$1 + 2 + 3 + \ldots + n = \frac{n(n + 1)}{2}.$$

Since $n(n + 1) = n^2 + n$, we see why the first formula is true. Archimedes discovered the second one by an ingenious method. Arab mathematicians around the year 1000 developed a geometric way to get all the other formulas, one by one.

In these last two chapters, we worked on finding the area under a curve. In Chapter 28 we found the slope of a curve. The two problems appear to have nothing to do with each other. Yet it turns out that if you knew how to find the slopes of curves, you could use the information to find areas under curves easily. This amazing fact lies at the heart of calculus.

If you look back at Chapter 28, where we found the slope of a curve, and at this chapter, where we found the area under a curve, you will notice the techniques are similar. To find the slope of a curve, we approximated a tangent line by secants or chords and then found how their slopes behaved as they got closer and closer to the tangent. To find the area under a curve, we approximated it by staircases and then found how these areas behaved as they looked more and more like the curved area. In both cases, we had to devise a special method.

By contrast, calculus develops a few tools that help solve all sorts of problems, in particular finding slopes of curves and curved areas. It dramatically reduces the need to concoct a fresh approach to each problem.

Anyone who has studied algebra and trigonometry is prepared to master calculus. It's the natural next step. Perhaps these chapters will tempt you to explore this wonderful creation.

Calculus is the study of changing quantities. For instance, think of the curve $y = x^2$ as the path of a comet. The tangent line at any point on the orbit shows the direction in which the comet is moving at that point. If the force of gravity suddenly vanished, the comet would "fly off on the tangent line."

Calculus finds speeds if you know the position of a moving object at all times. Conversely, it finds the position if you know the speeds at all times. More generally, if you know how fast some quantity is changing, it will find the total change over a period of time. Conversely, if you know how much there is at any time, it will tell you how rapidly the quantity is changing. That quantity could be the amount of a pollutant in a lake, the amount of gasoline refined, or the size of a bacterial population.

Though it was invented in the seventeenth century by Newton and G. W. Leibniz (1646–1716), calculus is a key to much of the science and technology of the twentieth century. As the historian Arnold Toynbee (1889–1975) wrote in his autobiography, *Experiences,*

> Looking back, I feel sure that I ought not to have been offered the choice [whether to study Greek or calculus] . . . calculus ought to have been compulsory for me. One ought, after all, to be initiated into the life of the world in which one is going to live. I was going to live in the Western World . . . and the calculus, like the full-rigged sailing ship, is . . . one of the characteristic expressions of the modern Western genius.

··· 31 ···

The Circle and All the Odd Numbers

There is a surprising connection between the number π and all the odd whole numbers:

$$\frac{\pi}{4} = 1 - \frac{1}{3} + \frac{1}{5} - \frac{1}{7} + \frac{1}{9} - \cdots$$

The additions and subtractions on the right-hand side of the equation go on forever. The more terms you use, the closer you get to $\pi/4$. This gives a way to compute π as accurately as you please without even drawing a circle. All you need is paper and pencil or a calculator. I was amazed by this formula when I first saw it, and, though I have been familiar with it for years, I am still amazed. What astonishes me is not only that there is such a formula, but that mere mortals discovered it and showed that it is true.

In the West, J. Gregory (1638–1675) obtained it in 1671 and Leibniz in 1673. However, it was familiar to the mathematicians of India as early as 1500, and it is the Indian argument that I present here. The argument, being longer than those in earlier chapters, requires more concentration. Perhaps, as you read, you may want to make notes of the key steps and redraw the diagrams (on a larger scale) for yourself.

First we gather the tools developed in earlier chapters that we need to use here. From Chapter 18, we use the fact that when r is a number between –1 and 1, then

$$\frac{1}{1 - r} = 1 + r + r^2 + r^3 + r^4 + \cdots$$

If we replace r by $-s$, we obtain, when s is between -1 and 1,

$$\frac{1}{1+s} = 1 - s + s^2 - s^3 + s^4 - \ldots,$$

which is the form we need in this chapter. (We will use it only when s is positive.)

From Chapter 30, we need the fact that as n increases

$$\frac{1^k + 2^k + \ldots + n^k}{n^{k+1}} \quad \text{approaches} \quad \frac{1}{k+1}.$$

Here k is a fixed whole number.

Actually, we use a closely related fact. There are n terms in the numerator. The last term is n^k. Since

$$\frac{n^k}{n^{k+1}} = \frac{1}{n},$$

and $\frac{1}{n}$ approaches 0 as n increases, we may delete n^k from the numerator and conclude that

$$\frac{1^k + 2^k + \ldots + (n-1)^k}{n^{k+1}} \quad \text{approaches} \quad \frac{1}{k+1},$$

as n increases. We need this information only when k is even. For instance, when k is 2, we have

$$\frac{1^2 + 2^2 + \ldots + (n-1)^2}{n^3} \quad \text{approaches} \quad \frac{1}{3}.$$

We also will use something familiar to anyone who has blown up a photo to a poster or projected a slide onto a screen. Namely, all lengths are magnified by the same factor. In particular, when a triangle is magnified, the larger triangle has the same angles as the smaller one, but the lengths of its three sides are proportional to the lengths of the three sides of the smaller one. In Figure 1, the smaller triangle has sides of lengths a, b, and c. The larger triangle has sides of lengths A, B, and C.

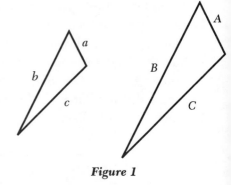

Figure 1

Then we have

$$\frac{a}{A} = \frac{b}{B} = \frac{c}{C}.$$

With all this machinery in place—the sum of a geometric series, the behavior of a certain quotient, and the property of magnification—we are ready to show that the ratio between the circumference of a circle and its diameter is related to all odd whole numbers. The reasoning has more steps than that in earlier chapters. However, they are all linked by one idea: calculating an estimate.

To begin, draw a (large) circle, or at least enough of it to show an angle of 45°, as in Figure 2.

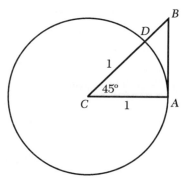

Figure 2

The circle has radius 1. Therefore, AB has length 1. The circumference of the circle is 2π. So the arc AD, which is one-eighth of the circumference, has length

$$\frac{2\pi}{8} = \frac{\pi}{4}.$$

So we already have $\pi/4$ shown geometrically, by an arc of a circle. All that remains is to see why the length of that arc is equal to $1 - \frac{1}{3} + \frac{1}{5} - \frac{1}{7} + \dots$.

We now use an approach that has already served us well in earlier chapters. We devise estimates of the arc length AD and then see how they behave as they get closer and closer to that length.

To form such estimates, first divide the line segment AB into many sections of equal length, as in Figure 3. Rays from C, the center of the circle, to the ends of these segments divide the arc AD into small pieces. While the little segments on the line AB are all the same size, the little arcs are not. (Those near *A* are longer than those near *D*. The higher the arc, the shorter it is.)

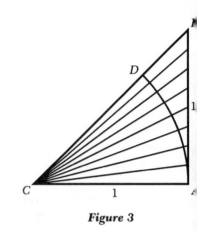

We then estimate the length of each of the small arcs into which we have cut the arc *AD*. The sum of these estimates is an estimate of the length of the arc *AD*, which we know is $\pi/4$. The more pieces that we cut the segment *AB* into, the better the estimates should be. Then we show that these estimates also approach $1 - \frac{1}{3} + \frac{1}{5} - \frac{1}{7} + \frac{1}{9} - \ldots$.

Figure 3

Now to the details. First we pick a whole number *n* and cut *AB* into *n* sections of equal length. Since *AB* has length 1, each of the *n* little segments has length $\frac{1}{n}$. Indirectly, with the aid of rays from the center of the circle, the arc *AD* is cut into *n* little arcs, but these arcs are not all the same size.

Let's take the case $n = 5$. That means that we cut *AB* into five equal segments, each of length $\frac{1}{5}$. The details of this case show what happens in the general case. I suggest that you make your own drawings, on a much larger scale than the figures you see here. That way, you can follow each step more easily. (Just as important, it also slows you to an appropriate pace.)

The picture for $n = 5$ is shown in Figure 4.

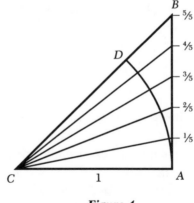

Figure 4

Now, let's estimate the length of the arc corresponding to the section from $\frac{3}{5}$ to $\frac{4}{5}$ on *AB*, as shown in Figure 5, where this portion of the arc is labeled *EF* and the segment from $\frac{3}{5}$ to $\frac{4}{5}$ is labeled *GH*.

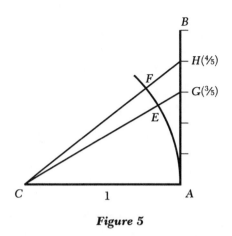

Figure 5

Draw *FI* tangent to the arc at *F*. Note that it is perpendicular to *CH*. *FI* is a good approximation to the arc *FE*, especially when *GH* is very small. So our goal will be to estimate *FI*. After all, it is easier to work with a straight line than with an arc of a circle.

To begin, draw *GJ* perpendicular to *CH*, as shown in Figure 6. Note that it produces two similar right triangles, *CFI* and *CJG*. We use that similarity later.

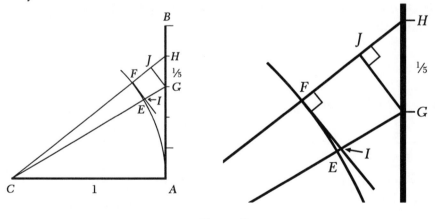

Figure 6

Note also that right triangles *GJH* and *CAH* are similar, since they share an angle at *H*. Using this similarity, we obtain

$$\frac{GJ}{AC} = \frac{GH}{CH}. \tag{1}$$

In equation (1), we know that *AC* is 1 and *GH* is ⅕. Thus

$$GJ = \frac{1/5}{CH}. \tag{2}$$

Before we find a formula for CH, let us use the other pair of similar triangles to get an equation involving the segment FI, which is our main interest.

By similar triangles CFI and CJG, we have

$$\frac{FI}{GJ} = \frac{CF}{CJ}. \tag{3}$$

Therefore,

$$FI = \frac{GJ \times CF}{CJ}. \tag{4}$$

Since CF is 1, this last equation for FI reduces to

$$FI = \frac{GJ}{CJ}. \tag{5}$$

What can we say about CJ? When n is large and GH is therefore small, so is JH. That tells us that CH is a good approximation to CJ in the sense that CH/CJ is close to 1. To make use of the fact that $CH/CJ \approx 1$ we rewrite equation (5) as

$$FI = \frac{GJ}{CJ} \times \frac{CH}{CH}. \tag{6}$$

Since CH/CJ is close to 1, we have

$$FI = \frac{GJ}{CH}. \tag{7}$$

Combining equation (2), which gives us an expression for GJ, with equation (7) leads to

$$FI \approx \frac{(1/5)/CH}{CH} = \frac{1/5}{CH^2}. \tag{8}$$

Now the time has come to treat CH. Since CH is the hypotenuse of right triangle CAH, we have

$$CH^2 = CA^2 + AH^2 = 1 + \left(\frac{4}{5}\right)^2. \tag{9}$$

Combining equations (8) and (9) provides this estimate of *FI*:

$$FI \approx \frac{1/5}{1 + (4/5)^2}. \tag{10}$$

Since *FI* is an estimate of the arc length *FE*, we finally have the estimate

$$FE \approx \frac{1/5}{1 + (4/5)^2}. \tag{11}$$

Take a close look at equation (11), for it reveals what happens in the general case. The numerator ⅕ will be replaced by ¹⁄ₙ. The ⅘ in the denominator will be replaced by a fraction that describes the length *AH*. But let's finish the case $n = 5$ first.

The estimate of the arc length *AD* when *AB* is cut into five segments is the sum of the estimates of five short arcs. Each estimate will look like the estimate in equation (11). In each, the numerator is ⅕. The denominator is different in each, going from $1 + (1/5)^2$ all the way to $1 + (5/5)^2$. All told, this is our estimate of the arc length *AD*:

$$\frac{1/5}{1 + (1/5)^2} + \frac{1/5}{1 + (2/5)^2} + \frac{1/5}{1 + (3/5)^2} + \frac{1/5}{1 + (4/5)^2} + \frac{1/5}{1 + (5/5)^2}. \tag{12}$$

The sum (12) is an estimate of the arc length *AD*, which we know is $\pi/4$. (Just out of curiosity, let's see how close this estimate is to $\pi/4$. To four decimal places, the sum is 0.7337 and $\pi/4$ is 0.7854. Not so bad, considering that we used only five segments and made some approximations in the process.)

The form of sum (12) shows what the sum will look like in the general case when 5 is replaced by *n*. There will be *n* summands instead of five summands. In each summand, the numerator will be ¹⁄ₙ instead of ⅕. The denominators will start at $1 + (1/n)^2$ and gradually go up to $1 + (n/n)^2$. For example, when *n* is 6, the estimate of the arc length *AD* is the sum

$$\frac{1/6}{1+(1/6)^2} + \frac{1/6}{1 + (2/6)^2} + \frac{1/6}{1 + (3/6)^2} + \frac{1/6}{1 + (4/6)^2} + \frac{1/6}{1 + (5/6)^2} + \frac{1/6}{1 + (6/6)^2}. \tag{13}$$

To four decimal places, this sum is 0.7426, a bit better estimate of $\pi/4$.

But we are really interested in what happens to estimates like (12) and (13) as the number of summands, *n*, increases. The general estimate, which has *n* summands, has the form

$$\frac{1/n}{1+(1/n)^2} + \frac{1/n}{1+(2/n)^2} + \ldots + \frac{1/n}{1+((n-1)/n)^2} + \frac{1/n}{1+(n/n)^2}. \quad (14)$$

In the sum (14) we show only the first two and the last two summands. The three dots between them stand for all the rest of the terms—the bulk of the sum.

What happens to the sum (14) as n gets larger and larger? To find out, we use the machinery described at the beginning of the chapter.

First of all, let's factor out $1/n$, which appears in every numerator in (14), rewriting (14) as

$$\frac{1}{n}\left[\frac{1}{1+(1/n)^2} + \frac{1}{1+(2/n)^2} + \ldots + \frac{1}{1+((n-1)/n)^2} + \frac{1}{1+(n/n)^2}\right]. \quad (15)$$

Let's work on the sum in the brackets.

Recall that if s is a positive number less than 1, then

$$\frac{1}{1+s} = 1 - s + s^2 - s^3 + s^4 - \ldots.$$

We may apply this formula to each summand in the brackets except the very last one (since n/n is not less than 1). In the first one, s is $(1/n)^2$; in the second, it is $(2/n)^2$. In the next to the last, it is $((n-1)/n)^2$. The last summand, $1/(1+(n/n)^2)$, equals $1/(1+1^2)$, which is $\frac{1}{2}$. But in the estimate (15), it is multiplied by $1/n$, so it contributes only $1/(2n)$ to the total estimate. When n increases, this contribution tends toward 0, and we will simply omit this part of the sum.

So we have

$$\frac{1}{1+(1/n)^2} = 1 - \left(\frac{1}{n}\right)^2 + \left(\frac{1}{n}\right)^4 - \left(\frac{1}{n}\right)^6 + \ldots$$

$$\frac{1}{1+(2/n)^2} = 1 - \left(\frac{2}{n}\right)^2 + \left(\frac{2}{n}\right)^4 - \left(\frac{2}{n}\right)^6 + \ldots \quad (16)$$

$$\vdots \qquad \vdots \qquad \vdots \qquad \vdots \qquad \vdots$$

$$\frac{1}{1+((n-1)/n)^2} = 1 - \left(\frac{n-1}{n}\right)^2 + \left(\frac{n-1}{n}\right)^4 - \left(\frac{n-1}{n}\right)^6 + \ldots.$$

Note that the +'s and the −'s alternate on the right side of each equation.

We add up these $n-1$ equations by first adding up the 1s, since the first term on the right is always 1. The sum of the 1s is $n-1$.

Next let's add up the second terms on the right side of (16). They add up to

$$-\left(\frac{1}{n}\right)^2 - \left(\frac{2}{n}\right)^2 - \ldots \left(\frac{n-1}{n}\right)^2.$$

Factoring n^2 out of all the denominators, we rewrite this sum as

$$-\frac{1^2 + 2^2 + \ldots + (n-1)^2}{n^2}.$$

Similarly, the third terms on the right of (16) add up to

$$\frac{1^4 + 2^4 + \ldots + (n-1)^4}{n^4}.$$

The other summands on the right of (16) lead to similar formulas.

With this information, we rewrite (15). Putting the factor $\frac{1}{n}$ that appears in the front of (15) back in to each summand, we see that (15) equals

$$\frac{n-1}{n} - \frac{1^2 + 2^2 + \ldots + (n-1)^2}{n^3} + \frac{1^4 + 2^4 + \ldots + (n-1)^4}{n^5} - \ldots . \qquad (17)$$

All that is left is to find out what happens to each summand in (17) as n increases.

The first term in the sum is

$$\frac{n-1}{n},$$

the ratio of two consecutive whole numbers. For instance, when n is 100, this ratio is $\frac{99}{100}$, which is 0.99. As n increases, the ratio $(n-1)/n$ approaches 1.

The second term in the sum is

$$-\frac{1^2 + 2^2 + \ldots (n-1)^2}{n^3}.$$

As we recalled in the machinery listed at the beginning of the chapter, this expression approaches $-\frac{1}{3}$ as n increases. In a similar manner, the third summand in (17) approaches $\frac{1}{5}$. The fourth summand, which is not shown in (17), approaches $-\frac{1}{7}$. All told, as n gets larger and larger, the expression (17) approaches

$$1 - \frac{1}{3} + \frac{1}{5} - \frac{1}{7} + \ldots$$

Recalling that these sums also approach an arc of length $\pi/4$, we see that

$$\frac{\pi}{4} = 1 - \frac{1}{3} + \frac{1}{5} - \frac{1}{7} - \ldots \qquad (18)$$

To see π expressed in terms of all the positive odd integers, multiply this last equation by 4, obtaining

$$\pi = 4\left(1 - \frac{1}{3} + \frac{1}{5} - \frac{1}{7} + \ldots\right).$$

The argument that gave us this equation may, on a first reading, seem a bit complicated. But if you go over it a few times, it should become simpler and simpler, for it is guided by just one idea: Make a general estimate and see how it behaves. One of the techniques was to add up a bunch of sums that were given as rows, left to right, by adding them up by columns, top to bottom. Mathematicians use this technique often.

There are other ways to establish (18). For instance, students in beginning calculus will obtain that formula by looking at the area under the curve $1/(1 + x^2)$ and above the interval from 0 to 1 in two different ways. After expressing $1/(1 + x^2)$ as a geometric series, they will apply the machinery of integral calculus. Their calculations will take only a few lines.

··· 32 ···

One Thought in Parting

Up to now I've exercised restraint, letting the truth and beauty of mathematics speak for themselves. Far be it from me to tell a reader that such and such a discovery or proof is beautiful. The discoveries themselves illustrate mathematical truth, absolutely certain and eternal, never to be rebutted by clever rhetoric or undercut by changing fashion.

Now, as I come to the end of the book, I'll present a proof that I openly assert is beautiful.

Recall that in Chapter 18 we showed that for a number r between -1 and 1, $1 + r + r^2 + r^3 + r^4 + \ldots = 1/(1 - r)$. We gave two different arguments for this. In one of them, we drew the points on the number line corresponding to $1, r, r^2, r^3, r^4, \ldots$ and then looked at the lengths of the little sections between them. In another, we just noticed that positive and negative terms in a certain sum cancel. In view of the importance of the sum of a geometric series, even the sum of just a front section, $1 + r + r^2 + \ldots + r^k$, it deserves a more beautiful and geometric proof. It certainly is important, since we used it to analyze banking in Chapter 19, to determine what happens to $(x^3 - 1)/(x - 1)$ when x is near 1 in Chapter 27, to find the slope of a curve in Chapter 28, to calculate the area under a curve in Chapter 30, and to obtain a formula for π in Chapter 31.

I don't know who invented this proof, which works when r is positive. (If r is negative, $r = -s$, where s is positive. Then $1 + r + r^2 + r^3 + r^4 + r^5 + r^6 + \ldots$ equals $1 - s + s^2 - s^3 + s^4 - s^5 + s^6 - \ldots$, which equals $(1 + s^2 + s^4 + s^6 + \ldots) - s(1 + s^2 + s^4 + \ldots)$. The geometric series in parentheses sums to $1/(1 - s^2)$.) My guess is that a mathematician came upon it by accident when looking into something else, but I don't know this for sure. In any case, it's part of the "folk literature" of mathematics, and it is quite short.

Begin by laying off segments of lengths $1, r, r^2, r^3, r^4, \ldots$ along a line, as in the following diagram:

Then at the left end of each section draw a vertical post as long as the section, obtaining the following diagram.

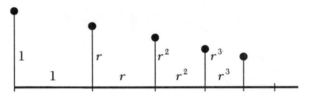

We are almost finished. The dots perched like lightbulbs on top of the posts lie on a straight line. (To check this, just compute the slopes of the segments joining each pair of adjacent dots. You will see that these slopes are all the same.)

Next we draw the line on which all those dots lie, as shown in the following diagram.

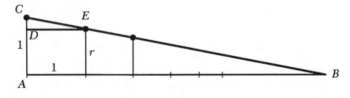

Already we see that the sum of all the powers of r does not get arbitrarily large, since their sum is just the length of the segment AB. All that remains is to find that length.

By similar triangles CAB and CDE, we have

$$\frac{AB}{DE} = \frac{AC}{DC}.$$

A glance at the diagram shows that $DE = 1$, $AC = 1$, and $DC = 1 - r$. Therefore,

$$\frac{AB}{1} = \frac{1}{1 - r},$$

hence,

$$AB = \frac{1}{1 - r}.$$

We conclude, in shorthand, that

$$1 + r + r^2 + r^3 + \ldots = \frac{1}{1 - r}.$$

You may complain, "Yes, it is rather neat, but what about the sum of just a finite number of terms, for instance, the sum $1 + r + r^2$?" A smaller triangle similar to ABC will show that this sum equals $(1 - r^3)/(1 - r)$, as you may check.

Maybe not everyone will agree with me, for beauty is a matter of taste, but I think this third proof is indeed beautiful: It is visual, it is short, and, once it is called to one's attention, it seems as inevitable and memorable as a Mozart symphony, as if it had been waiting in heaven from the beginning of time for someone to bring it down to earth. If God keeps a book of proofs, this is one that I think would be found in its pages.

Note that in this proof geometry sheds light on algebra. In the chapters on slope and area under a curve, algebra sheds light on geometry. This is typical of mathematics, where seemingly unrelated fields often connect in surprising and delightful ways.

Having used similar triangles just now to sum a geometric series, I can't resist showing how to use them to obtain a proof of the Pythagorean theorem different from the one in Chapter 22. This is another proof that I find charming. It goes like this.

Consider a right triangle with the sides of the usual lengths a, b, and c, with a less than or equal to b. Draw the line from the corner of the right angle perpendicular to the hypotenuse. This line divides the big triangle into two smaller triangles, and the hypotenuse into two segments, of lengths d and e, with $d + e = c$, as shown in this diagram.

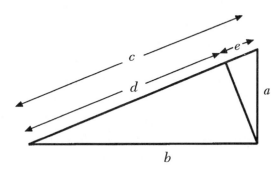

Though the big right triangle and the two smaller ones are not the same size, they have the same angles, as you may check. Therefore, they are similar. That is the key to the proof.

Since the right triangle whose hypotenuse is a is similar to the triangle whose hypotenuse is c, we have

$$\frac{e}{a} = \frac{a}{c}.$$

On the left side of this equation are the smallest sides of the two triangles; on the right side are the hypotenuses.

Multiplying both sides of this equation by ac gives us the equation

$$ce = a^2.$$

The same reasoning, using the triangle whose hypotenuse is b and the triangle whose hypotenuse is c, gives us

$$cd = b^2.$$

Adding the equations $ce = a^2$ and $cd = b^2$ tells us that

$$cd + ce = a^2 + b^2.$$

By that great link between addition and multiplication, the distributive law, we have $cd + ce = c(d + e)$. Thus

$$c(d + e) = a^2 + b^2.$$

Recalling that $d + e = c$, we have obtained the Pythagorean theorem:

$$c^2 = a^2 + b^2.$$

There are two things I like about this proof. First, it uses only objects inside the triangle—not three squares outside of it. Second, it uses only lengths of lines rather than areas. (Area is a much more complex idea than length.) On the other hand, geometers may complain, "There's too much algebra in this. I prefer the proof, using three squares, in Chapter 22, for it's totally pictorial."

This just goes to show that taste matters in mathematics, as it does in music and art. Even if everyone agrees that two different proofs are both correct, there is ample room for healthy disputes over whether one proof is more beautiful than the other.

With the detour of this little chapter, we come to the end of our journey. In this book we have gazed upon some of the eternal truths and beauties of mathematics, encountered astonishing discoveries, and followed each of the steps in the logic that leads to them. But mathematics is not simply a self-contained alternative universe. The world of daily life, of business and industry, freely makes use of its language and its techniques.

The path has taken us only a little way into the kingdom, a realm full of countless beautiful theorems and challenging mysteries. It promises anyone willing to explore it—however practical or ethereal in temperament—endless delights, far more than any lifetime can encompass.

FURTHER READING

When I was a guest on a radio talk show, a listener called in for advice: "I would like to know more mathematics. What do you suggest I read?" This was a question I had never considered. One reason I didn't give a satisfactory response is that the suggestion depends on the level of mathematics the listener has already reached. Now, somewhat belated but perhaps of use to others, is the answer I should have given then.

If you want to study the basic ideas of arithmetic, browse in a bookstore through the many paperback books that prepare people to take entrance examinations. Here are a few examples:

> Carman, R., and M. Carman. *Quick Arithmetic: A Self-Teaching Guide*, 2d ed. New York: John Wiley & Sons, 1984.

> Goozner, C. *Business Mathematics the Easy Way*, 2d ed. Hauppauge, NY: Barron's Educational Series, 1991.

> Sia, A., et al. *Entry Level Math*. Piscataway, NJ: Research and Education Association, 1993.

> Slavin, S. *All the Math You'll Ever Need*. New York: John Wiley & Sons, 1989.

Certain of these cover some algebra and geometry. It is surprising how few pages are needed to present the essential ideas and techniques.

If you have a good grasp of arithmetic, especially of fractions, and want to study algebra, you might turn to Haym Kruglak and J. T. Moore, *Basic Mathematics with Applications* (New York: Schaum's Outline Series, McGraw-Hill, 1973). The first ten chapters cover arithmetic and algebra, with the remaining chapters treating trigonometry and a few other topics needed for calculus. Chapters 13 and 16 on trigonometry take only 27 pages, but they present the essential information. P. Selby and S. Slavin, *Practical Algebra; A Self-Teaching Guide*, 2nd ed. (New York: John Wiley & Sons, 1991), covers basic algebra and includes many self-tests.

If you studied algebra some time ago and want to brush up, try P. Selby and S. Slavin, *Quick Algebra Review: A Self-Teaching Guide* (New York: John Wiley & Sons, 1993).

S. Slavin, *Quick Business Math* (New York: John Wiley & Sons, 1995), begins with the rudiments of algebra and percentages, and then applies them to retailing, simple and compound interest, depreciation, and statistics.

If you know algebra and want to prepare to study calculus, you might look at J. Stewart, L. Redlin, and S. Watson, *Mathematics for Calculus* (Pacific Grove, CA: Brooks Cole, 1992), or D. Cohen, *Precalculus*, 4th ed. (Minneapolis: West, 1993).

However, with a sound command of algebra, geometry, and trigonometry, you could begin calculus. The three books of the Guided Inquiry Series that I wrote with D. Chakerian and C. Crabill cover algebra, geometry, and trigonometry. Though written for small-group study, they have been effective even with people who work through them alone, for instance, in a home-study program. They are published by Sunburst, 101 Castleton Street, Pleasantville, New York, 10570-3498 (800-338-3457).

You can choose from many calculus books. My own, *Calculus and Analytic Geometry*, 5th ed., written with Anthony Barcellos and published by McGraw-Hill, New York, I know is suitable for self-study. It includes a short section on the necessary trigonometry. You can usually pick up an inexpensive copy of an earlier edition of it or of other calculus books in a secondhand store.

If you are looking for ways to interest children in mathematics, consult L. Polonsky, et al., *Math for the Very Young: A Handbook for Parents and Teachers* (New York: John Wiley & Sons, 1995). It presents many ways to introduce children (even of preschool age) to a variety of mathematical concepts, using familiar objects and activities around the home, in the zoo, and so on.

To encourage girls to study mathematics and science, you might obtain a copy of M. Parker, ed., *She Does Math—Real-Life Problems from Women on the Job* (Washington, DC: Mathematical Association of America, 1995). The problems, which are typical of those met in the real world, use only high school mathematics. The book reviews the careers of 38 professional women and some of the problems they worked on.

If you want to sample other types of mathematics, take a look at any of the following books:

Beckmann, P. *A History of Pi*. New York: St. Martin's Press, 1971.

Courant, R., and H. Robbins. *What Is Mathematics?* New York: Oxford University Press, 1960.

Dunham, W. *Journey through Genius*. New York: John Wiley & Sons, 1990. This book presents 12 theorems, placing their proofs in a personal and historical context.

———. *The Mathematical Universe*. New York: John Wiley & Sons, 1994.

Huff, D. *How to Lie with Statistics.* New York: Norton, 1993. This classic is accessible to readers with very little mathematics background.

Maor, E. *e: The Story of a Number.* Princeton, NJ: Princeton University Press, 1993.

Paulos, J. *Innumeracy, Mathematical Illiteracy and Its Consequences.* New York: Vintage Books, 1986. This title explores the arithmetic of a variety of topics, such as parapsychology, the stock market, coincidences, pseudoscience, drug testing, and so on.

Steen L., ed. *Mathematics Today.* New York: Springer-Verlag, 1978. Twelve informal essays.

———, ed. *For All Practical Purposes.* New York: W. H. Freeman, 1988.

Stein S. *Mathematics: The Man-made Universe,* 3d ed. New York: McGraw-Hill, 1976.

For the history of mathematics:

Boyer, C. B., and U. C. Merzbach. *A History of Mathematics.* New York: John Wiley & Sons, 1991.

Edwards, C. H., Jr. *The Historical Development of the Calculus.* New York: Springer-Verlag, 1979.

Eves, H. W. *An Introduction to the History of Mathematics with Cultural Connections.* Philadelphia: W. B. Saunders, 1990.

Kline, M. *Mathematics in Western Culture.* New York: Oxford University Press, 1953.

———. *Mathematical Thought from Ancient to Modern Times.* New York: Oxford University Press, 1972.

For biographies:

Albers, D. J., and G. L. Alexanderson. *Mathematical People: Profiles and Interviews.* Boston: Birkhäuser, 1985.

Bell, E. T. *Men of Mathematics.* New York: Simon and Schuster, 1935.

These are only a few of the many books available. For a far more extensive list, consult L. Steen, ed., *Library Recommendations for Undergraduate Mathematics* (1529 18th Street, NW, Washington, DC 20036: Mathematical Association of America, 1992).

GLOSSARY OF SYMBOLS

The page on which the symbol is introduced is given at right.

REFERENCES

1. The Many Faces of Mathematics

Sources

4   **"I'm not out to convince":** Conrad Hilton, *Be My Guest* (Englewood Cliffs, NJ: Prentice Hall, 1957), 67.

4   **"First there's mathematics":** Charles Munger, "A Lesson on Elementary, Worldly Wisdom As It Relates to Investment Management and Business," *Outstanding Investor Digest* 10, nos. 1 and 2 (May 1995): 5.

6   **"Those who talk of money":** John Kenneth Galbraith, *Money, Whence It Came, Where It Went* (Boston: Houghton Mifflin, 1975), 4.

6   **"The important thing":** Richard Dawkins, *The Blind Watchmaker* (New York: Norton, 1986), 67.

7   **"I must study politics":** John Adams, *Letters of John Adams, Addressed to His Wife,* vol. 2 (Boston: Phillips and Sampson, 1848), 68.

9   **"Having to conduct my grandson":** Thomas Jefferson, *Jefferson Himself,* ed. Bernard Mayo (Charlottesville: University Press of Virginia, 1970), 293.

2. The Spell of Cool Numbers

Sources

11   **description of the Thirteen Club:** William Walsh, *Handy-book of Literary Criticism* (Philadelphia: Lippincott, 1892), 1051.

13   **"Why a cease-fire":** M. Gordon and B. Trainor, "How Iraq Escaped to Threaten Kuwait Again," *New York Times,* 23 October 1994.

3. Hot Numbers

Sources

15   **"The more details the negotiator":** Joshua Stein, "The Art of Real Estate Negotiations," *Real Estate Review* (Winter 1996): 48–53.

17   **discussion of federal deficit:** Michael Wines, "Whose Deficit, and How Big?" *New York Times,* 30 October 1994.

17   **"This year's model costs":** James Bennet, "Will Rising Prices of Cars Imperil Detroit's Recovery?" *New York Times,* 22 August 1994.

17   **discussion of education:** Albert Shanker, "Where We Stand," *New York Times,* 28 May 1995.

18   **"If we chose the top 3%":** Howard Wainer, "Does Spending Money on Education Help?" *Educational Researcher* 22 (1993): 22–24.

    See also David Berliner and Bruce Biddle, *The Manufactured Crisis* (Reading, MA: Addison-Wesley, 1995) for a discussion of other hot numbers in education.

19   **"This is not true":** Humphrey Taylor, "Two Margins of Error," Letter to the Editor, *New York Times,* 4 November 1994.

19   **"So many people have so much invested":** Lawrence K. Altman, "Obstacle-Strewn Road to Rethinking the Numbers on AIDS," *New York Times,* 1 March 1994.

4. Don't Do a Number on Me

Sources

22   **"There are at least fifty ways":** Harry Hopkins, *The Numbers Game* (London: Secker and Warburg, 1973), 245.

22   **"parents' putting pressure":** Daniel Goleman, "Seventy-five Years Later, Study Is Still Tracking Geniuses," *New York Times*, 7 March 1995.

22   **"will power, perseverance":** Ibid.

      See also Jim Holt, "Anti-Social Science?" *New York Times*, 19 October 1994, and Stephen Jay Gould, *The Mismeasure of Man* (New York: Norton, 1981).

5. Anecdote versus Number

Sources

26   **"I see some kind of hospital":** N. Klyver and M. Reiser, "A Comparison of Psychics, Detectives, and Students in the Investigation of Major Crimes" (Los Angeles Police Department, 150 North Los Angeles Street, Los Angeles, CA 90012).

      See also M. Reiser et al., "An Evaluation of the Use of Psychics in the Investigation of Major Crimes," *Journal of Police Science and Administration* 7 (1979): 18–25, and E. M. Hansel, *ESP: A Scientific Evaluation* (New York: Scribners, 1966).

29   **study of driving accidents:** S. Stein, "Risk Factors of Sober and Drunk Drivers by Time of Day," *Alcohol, Drugs, and Driving* 5 (1989): 215–227.

6. It Ain't Necessarily So

Sources

36   **"No mathematician should ever":** G. H. Hardy, *A Mathematician's Apology* (New York: Cambridge University Press, 1967), 70.

37   **"To make a right triangle":** Arthur Coxford, Zalman Usiskin, and Daniel Hirschhorn, eds., *University of Chicago Geometry* (Glenview, IL: Scott, Foresman, 1993), 393.

38   **"The Egyptians had carpenter squares":** Moritz Cantor, *Vorlesungen über Geschichte der Mathematik*, vol. 1 (Leipzig: Teubner, 1907), 105.

39   **discussion of the golden ratio:** George Markowsky, "Misconceptions about the Golden Ratio," *The College Mathematics Journal* 23 (1992): 2–19.

40   **"When he went down to the bathing pool":** Vitruvius, *On Architecture*, vol. 2 (Cambridge: Harvard University Press, 1956), 205.

40   **"Archimedes declared":** Plutarch, *Plutarch's Lives*, vol. 5, trans. Bernadotte Perin (New York: G. P. Putnam's Sons, 1917), 473.

41   **"All night he had spent":** E. T. Bell, *Men of Mathematics* (New York: Simon and Schuster, 1937), 375.

41   **"Galois had been submitting papers":** Tony Rothman, "Genius and Biographers: The Fictionalization of Evariste Galois," *American Mathematical Monthly* 89 (1982): 84–106.

42   **"He found that the sum":** Morris Kline, *Mathematical Thought from Ancient to Modern Times* (New York: Oxford University Press, 1972), 872–873.

42   **Gauss's experiment and Einstein's theory of relativity:** Arthur I. Miller,
      "The Myth of Gauss' Experiment on the Euclidean Nature of Physical Space,"
      Isis 63 (1972): 345–348.

42   **the myth of Einstein's weakness in arithmetic:** W. Sullivan, "Einstein
      Revealed as Brilliant as Youth," *New York Times,* 14 February, 1984.

42   **The Nobel Prize and mathematics:** Lars Gårding and Lars Hörmander,
      "Why Is There No Nobel Prize in Mathematics?" *Mathematical Intelligencer*
      7, no. 3 (1985): 73–74.

7. The Rapid Idiots

Sources

46   **"A computer certainly can help you":** Joshua Stein, "How to Prevent
      Mistakes in Transactional Legal Work," *The Practical Lawyer* 40 (1994): 51–64.

47   **"Although technology has advanced":** M. J. and J. J. Flannery, "Causes of
      Hotel Industry Distress," *Real Estate Review* 20 (Fall 1990): 35–39.

8. The Mother of Invention

Source

53   **"No one has yet discovered":** Hardy, *A Mathematician's Apology,* 140.

For Further Reading

On knots: Colin Adams, *The Knot Book* (New York: W. H. Freeman, 1994). This
      book describes in detail the relation of knots to DNA.

        G. Kolata, "Solving Knotty Problems in Mathematics and Biology," *Science*
        231 (1986): 1506–1508.

On Radon: Stanley R. Deans, *The Radon Transform and Some of Its Applications*
      (New York: John Wiley & Sons, 1983). The first chapter describes applica-
      tions to medicine, astronomy, microscopy, optics, and geophysics.

On codes: G. Kolata, "100 Quadrillion Calculations Later, Eureka!" *New York
      Times*, 27 April 1994.

        Martin Hellman, "The Mathematics of Public-key Cryptography," *Scientific
        American* (August 1979): 146–157.

        R. L. Rivest, A. Shamir, and L. Adleman, "A Method for Obtaining Digital
        Signatures and Public-Key Cryptosystems," *Communications of the
        Association for Computing Machinery* 21 (1978): 120–126.

        G. Kolata, "Hitting the High Spots of Computer Theory," *New York Times*,
        13 December 1994.

On other applications: M. R. Schroeder, *Number Theory in Science and
      Communication*, 2d ed. (New York: Springer-Verlag, 1986). This book
      describes various applications of number theory, including secret codes.

        Eugene P. Wigner, "The Unreasonable Effectiveness of Mathematics
        in the Natural Sciences," *Communications on Pure and Applied
        Mathematics* 13 (1960): 1–14. This is an often reprinted survey of some
        surprising applications of mathematics.

        F. E. Browder, "The Relevance of Mathematics," *American Mathematical
        Monthly* 83 (1976): 149–154.

9. What Is a Job, Really?

Sources

58 **"almost one in three":** Robin Wright, "A Revolution at Work," *Los Angeles Times*, 7 March 1995, World Report.

58 **"Why are only movie stars":** Richard M. Smith, "And Now, We'd Like to Bring You Some Good News," *Newsweek*, 29 May 1995.

10. What's in It for Me?

Sources

61 **"Many high school students":** "Survey Finds Students Uninformed about Need for Mathematics," *Notices of the American Mathematics Society* 42 (November 1995): 1299–1300.

62 **"In my career":** Deborah L. Jacobs, "Stepping into the Heels of Their Career Moms," *New York Times*, 8 May 1994.

62 **"which of some 60":** Hal Saunders, preface to *When Are We Ever Going to Use This?* (Palo Alto, CA: Dale Seymour Publications, 1988).

69 **"Mathematics opened the doors":** *Mathematical Scientists at Work*, 2d ed. (Washington, DC: Mathematical Association of America, 1993).

69 **"Although I had no background":** Ibid.

69 **"Mathematics [is] so useful":** Jefferson, *Jefferson Himself*, 17.

70 **"Deciding how much high school math":** Hall Dillon, "Math and Your Career," *Occupational Outlook Quarterly,* 31, no. 2 (Summer 1993): 27–29.

For Further Reading

U. S. Department of Labor, Employment and Training Administration, *Dictionary of Occupational Titles*, 4th ed., 2 vols. (Lanham, MD: Bernan Press, 1991).

Bureau of Labor Statistics, *Occupational Outlook Handbook,* an annual.

Bureau of Labor Statistics, *Occupational Outlook Quarterly.*

"Job-Related Education and Training: Their Impact on Earnings," *Monthly Labor Review* (October 1993): 21–31.

Les Krantz, *The Jobs Rated Almanac* (New York: World Almanac, 1992). This ranks 250 jobs by such factors as security, benefits, stress, income, outlook, physical demands, and travel opportunities.

J. W. Wright and E. J. Dwyer, *The American Almanac of Jobs and Salaries* (New York: Avon Books, 1990).

William E. Hopke, *Encyclopedia of Careers and Vocational Guidance*, 9th ed. (Chicago: J. G. Ferguson Publishing Co., 1993).

J. Michael Farr, ed., *The Complete Guide for Occupational Exploration* (Indianapolis: JIST Works, 1993).

Bob Adams, Inc. of Holbrook, MA, and Survey Books of Chicago, IL, each puts out a series of books describing the job opportunities in specific cities and states.

Peggy Kneffel Daniels and Susan E. Edgar, eds., *Job Seeker's Guide to Private and Public Companies,* 2d ed. (Detroit: Gale Research Inc., 1994). This profiles more than 17,000 companies.

11. The Action Syndrome

Sources

72 **"Artabanus said, 'It is best for men'":** Manuel Komroff, ed., *The History of Herodotus* (New York: Tudor, 1947), 373–375.

72 **"There are two parts to the human mind":** "Champion Body Builder," *Sacramento Bee,* 22 October 1978.

72 **"When making a decision":** Theodor Reik, *Listening with the Third Ear* (New York: Farrar, Straus, 1948), vii.

72 **"The essence of ultimate decision":** Graham T. Allison, *The Essence of Decision* (Boston: Little, Brown and Co., 1971), quoted in the front matter.

72 **"It started as a pilot project":** "Continental Is Dropping 'Lite' Service," *New York Times,* 14 April 1995.

12. Where Have All the Reforms Gone?

Sources

75 **"One of the most obvious facts":** N. J. Lennes, "Modern Tendencies in the Teaching of Algebra," *The Mathematics Teacher* 1 (1908): 94–104.

75 **"Our conference is charged with gloom":** C. F. Wheelock, "Is the Average Secondary Pupil Able to Acquire a Thorough Knowledge of all the Mathematics Ordinarily Given in Their Schools?" *The Mathematics Teacher* 3 (1911): 101–128, in particular, 123–124.

75 **"The traditional curriculum":** Morris Kline, "The Ancients versus the Moderns, a New Battle of the Books," *The Mathematics Teacher* 51 (1958): 418–427, in particular, 424.

76 **"The student today":** Zalman Usiskin, Max Bell, Sheila Sconiers, Sharon Senk, and Izaak Wirszup, *The University of Chicago School Mathematics Project* (Autumn 1994): 4.

76 **"All the modern physical equipment":** I. J. Schwatt, "On the Curriculum of Mathematics," *The Mathematics Teacher* 3 (1910): 1–8, in particular, 6.

76 **"the distractions of modern life":** E. R. Smith, in Wheelock, "Is the Average Secondary Pupil Able?" 123–124.

76 **"Before the United States can":** Harold W. Stevenson and James W. Stigler, *The Learning Gap: Why Our Schools Are Failing and What We Can Learn from Japanese and Chinese Education* (New York: Summit Books, 1992), 201.

77 **"To start a reform":** W. H. Metzler, "Mathematics for Training and Culture," *The Mathematics Teacher* 2 (1909): 47–56, in particular, 47.

77 **"It is nonsense":** L. P. Benezet, "The Story of an Experiment," *Journal of the National Education Association* 24 (1935): 241–244, 301–303, in particular, 241; 25 (1936): 7–8.

77   **"represented my real belief":** Ibid., 241.

77   **"not one parent in ten":** Ibid.

77   **"the 6th graders were divided":** Ibid., 244.

78   **"why a correct answer":** Ibid., 303.

79   **"The separation of the [peaceful] earth satellite":** Dwight Eisenhower, *Waging Peace 1956–1961* (Garden City, NY: Doubleday, 1965), 209.

80   **"No one can predict exactly":** E. G. Begle, "The School Mathematics Study Group," *The Mathematics Teacher* 51 (1958): 616–618.

80   **"SMSG will combine":** Ibid., 617.

80   **"Curriculum reforms have been advocated before":** Burton W. Jones, "Silken Slippers and Hobnailed Boots," *The Mathematics Teacher* 52 (1959): 322–327, in particular, 326.

80   **"Those of us who are familiar":** James H. Zant, "Improving the Program in Mathematics in Oklahoma Schools," *The Mathematics Teacher* 54 (1961): 594–599.

81   **"Fellow teachers":** Wallace Manheimer, "Some Heretical Thoughts from an Orthodox Teacher," *The Mathematics Teacher* 53 (1960): 22–26.

81   **"It would be a tragedy":** "On the Mathematics Curriculum in the High School," *The Mathematics Teacher* 55 (1962): 191–195, in particular, 191.

81   **"Parents are discovering":** E. Begle, ed., preface to *A Very Short Course in Mathematics for Parents* (Palo Alto, CA: Stanford University Press, 1963).

84   **"In the early sixties my assistant":** Richard Feynman, *Surely You're Joking, Mr. Feynman* (New York: Bantam, 1985), 264–266.

84   **"The reception thus far":** William Wooton, *SMSG: The Making of a Curriculum* (New Haven: Yale University Press, 1965), 135.

86   **"To establish a broad framework":** *Curriculum and Evaluation Standards for School Mathematics* (Reston, VA: National Council of Teachers of Mathematics, 1989), v.

86   **"Students should value mathematics":** *Professional Standards for Teaching Mathematics* (Reston, VA: National Council of Teachers of Mathematics, 1991), 5.

86   **"Effective teachers can stimulate":** Ibid., 2. This is quoted from *Everybody Counts* (National Research Council, 1989), 58–59.

86   **"Some proficiency with paper and pencil":** *Curriculum and Evaluation Standards*, 8.

87   **"The calculator renders obsolete":** *Professional Standards*, 56.

87   **"The success of the current reform movement":** Jack Price, "Reform Is a Journey, Not a Destination," *NCTM News Bulletin* (November 1994): 3.

87   **"The New Math was totally abstract":** Elaine Rosenfeld, chair of the Mathematics Subject Matter Committee, Statement before the Curriculum Development and Supplemental Materials Commission, California Board of Education, 13 October 1994.

88   **"Even in the faddy world":** Chester Finn, Jr., "What if Those Math Standards Are Wrong?" *Education Week* 12 (20 January 1993): 36.

88   **"Dissent from the *Standards*":** Zalcman Usiskin, "What Changes Should Be Made for the 2nd Edition of the NCTM Standards?" *Humanistic Mathematics Network Journal* 10 (August 1994): 13–38.

88   **"Students must deal with whole situations":** James K. Bidwell and Robert G. Clason, *Readings in the History of Mathematics Education* (Washington, DC: National Council of Teachers of Mathematics, 1970), 538–566, in particular, 542, 553. Originally published in *Mathematics in General Education* (New York: D. Appleton-Century Co., 1938).

88   **"There is a trend":** Ibid., 586–617, in particular, 594. Originally published in *The Place of Mathematics in Secondary Education* (New York: Teachers College, Columbia University, 1940).

89   **"The ideal classroom climate":** *Mathematics Framework for California Public Schools* (Sacramento: California Department of Education, 1972), 7.

89   **"Begin with a paper triangle":** *Middle Grades Mathematics, Course 2* (Englewood Cliffs, NJ: Prentice Hall, 1995), 55–56.

92   **"The Japanese books":** Richard E. Mayer, Valerie Sims, and Hidetsugu Tajika, "A Comparison of How Textbooks Teach Mathematical Problem Solving in Japan and in the United States," *American Educational Research Journal* 32 (1995): 443–460, in particular, 456.

For Further Reading

J. K. Bidwell and R. G. Clason, *Readings in the History of Mathematics Education* (Washington, DC: National Council of Teachers of Mathematics, 1970).

Morris Kline, *Why Johnny Can't Add: The Failure of the New Math* (New York: St. Martin's Press, 1973).

———, *Why the Professor Can't Teach* (New York: St. Martin's Press, 1976).

S. K. Stein, "Gresham's Law: Algorithm Drives Out Thought," *Journal of Mathematical Behavior* 7 (1988): 79–84.

13. Some Proposals, Modest and Immodest

Sources

95   **"We never had any intention":** "Rejecting Barbie, Doll Maker Gains," *New York Times*, 1 September 1993.

96   **"A nation that is falling behind":** Stevenson and Stigler, *The Learning Gap*, 88.

96   **"The single most important activity":** "Read Aloud to the Kids," *Smithsonian* 25, no. 11 (February 1995): 82–91.

96   **"Rather than becoming disengaged":** Stevenson and Stigler, *The Learning Gap*, 217.

97   **"By the time":** David Hatchett, "Black College Athlete," *The CRISIS* 98, no. 9 (November 1991): 12–14.

97   **"The typical Japanese family":** Stevenson and Stigler, *The Learning Gap*, 54.

98   **"Chinese and Japanese children know":** Ibid., 68.

99   **"A mom's attitude":** Barbara F. Meltz, "For Girls, Comfort with Math Calls for Strong Maternal Role," *The Boston Globe*, 28 September 1995.

100 **"I thought the writing"**: Benjamin Franklin, *Benjamin Franklin's Auto-biography* (New York: Norton, 1984), 11–12.

101 **"In the current culture"**: *Professional Standards,* 185.

101 **"Complaints about the mathematical preparation"**: T. W. Hungerford, "Future Elementary Teachers: The Neglected Constituency," *The American Mathematical Monthly* 101 (1994): 15–21.

See also, "Research Mathematicians in Mathematics Education," *Notices of the American Mathematical Society* 35 (1988): 790–794, 1123–1131.

101 **"The mathematical preparation"**: quoted in Hungerford, *Future Elementary Teachers,* 15.

102 **"At least in the case of elementary school teachers"**: Luther Jennings, Stephen George, Anne Schell, *Professional Teaching Expertise: Fact or Myth* (manuscript).

103 **"Decrease the teaching load"**: Stevenson and Stigler, *The Learning Gap,* 207.

14. How to Read Mathematics

Sources

108 **"Buy the most elementary"**: Marilyn vos Savant, Ask Marilyn, *Parade Magazine,* 22 May 1994.

15. You Will Never See a Large Number

Sources

113 **googol discussion:** Edward Kasner and James Newman, *Mathematics and the Imagination* (New York: Simon and Schuster, 1940), 20–25.

113 **"There are some, King Gelon"**: Archimedes, *The Sand-Reckoner,* Great Books of the Western World (Chicago: Encyclopedia Britannica, 1952), 520.

118 **"Our proof is indirect"**: A. M. Odlyzko and H. J. J. te Riele, "Disproof of the Mertens Conjecture," *Journal für die reine und angewandte Mathematik* 357 (1985): 138–160.

16. The Car and Two Goats

Sources

121 **"Suppose you're on a game show"**: Marilyn vos Savant, "Ask Marilyn," *Parade Magazine,* 9 September 1990.

23. Pi Is a Piece of Cake—or Is It?

Source

173 **"after the class collects"**: Gina Kolata, "Math Is Only New When the Teacher Doesn't Get It," *New York Times,* 2 April 1989.

26. A Fresh Look at Kindergarten

Sources

198 **"May I ask you a question"**: Sherman K. Stein, *Mathematics: The Man-made Universe* (New York: McGraw Hill, 1994), 452.

199 **"I proposed my question"**: Ibid., 452.

30. Finding a Curved Area

Source

234 **"Looking back, I feel sure":** Arnold Toynbee, *Experiences* (New York: Oxford University Press, 1969), 12–13.

31. The Circle and All the Odd Numbers

For Further Reading

C. T. Rajagopal, "A Neglected Chapter of Hindu Mathematics," *Scripta Mathematica* 15 (1949): 201–209.

C. T. Rajagopal and T. V. Vedamurthi, "On the Hindu Proof of Gregory's Series," *Scripta Mathematica* 17 (1951): 65–74.

C. T. Rajagopal and M. S. Rangachari, "On an Untapped Source of Medieval Keralese Mathematics," *Archive for History of Exact Sciences* 18 (1977): 89–102.

INDEX